MW01285669

Nature is your starting point.
You have eyes to see,
nerves to feel,
a mind to understand.
Have respect for nature
and your natural responses.
Study the things that interest you,
that awaken your imagination
and nature will keep you sound.

~ Maynard Dixon

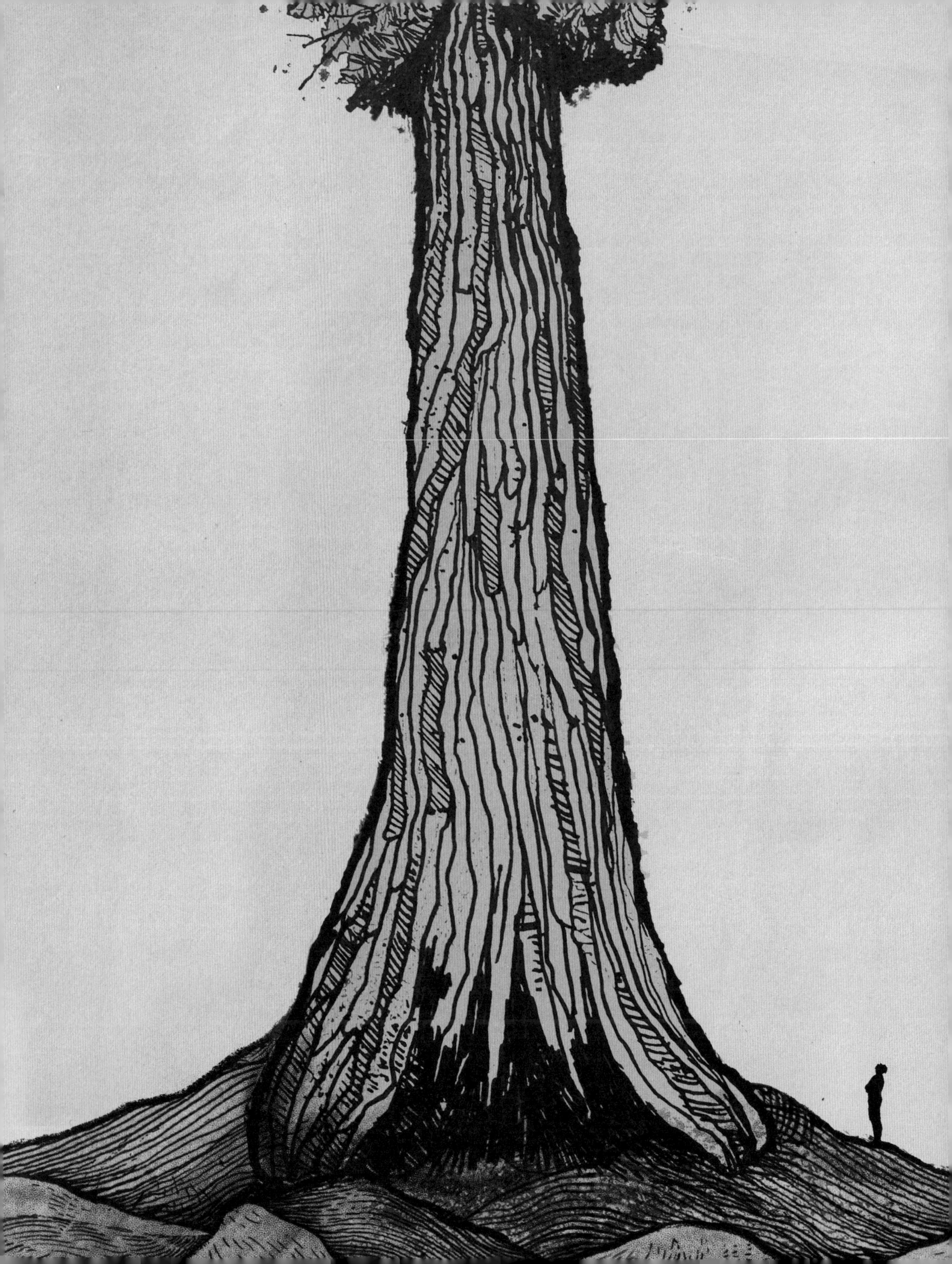

EVENTUALLY A SEQUOIA

STORIES OF ART, ADVENTURE & THE WISDOM OF GIANTS

JEREMY COLLINS

MOUNTAINEERS BOOKS

Seattle, USA

MOUNTAINEERS BOOKS is dedicated to the exploration, preservation, and enjoyment of outdoor and wilderness areas.

1001 SW Klickitat Way, Suite 201, Seattle, WA 98134
800-553-4453, www.mountaineersbooks.org

Printed in China
Distributed in the United Kingdom by Cordee, www.cordee.co.uk

28 27 26 25   1 2 3 4 5

Book design, layout, and art: Jeremy Collins
Production: Jen Grable
All photographs by the author except the following: pages 74, 87, 90, 94, 97, 99, 144, 148, 155, 162, 165 by James Q. Martin; pages 19, 42, 45, 48, 57 by Michael Clark; pages 9, 224 by Michael Paul Jones; page 140 by Devaki Murch

Library of Congress Cataloging-in-Publication data is on file for this title at https://lccn.loc.gov/2025932454

Mountaineers Books titles may be purchased for corporate, educational, or other promotional sales, and our authors are available for a wide range of events. For information on special discounts or booking an author, contact our customer service at 800-553-4453 or mbooks@mountaineersbooks.org.

Printed on FSC®-certified materials

ISBN: 978-1-68051-805-4
ISBN (Special Collector's Edition): 978-1-68051-854-2

An independent nonprofit publisher since 1960

© copyright 2025

Dedicated to Mom & Dad
(Joyce & Gary)

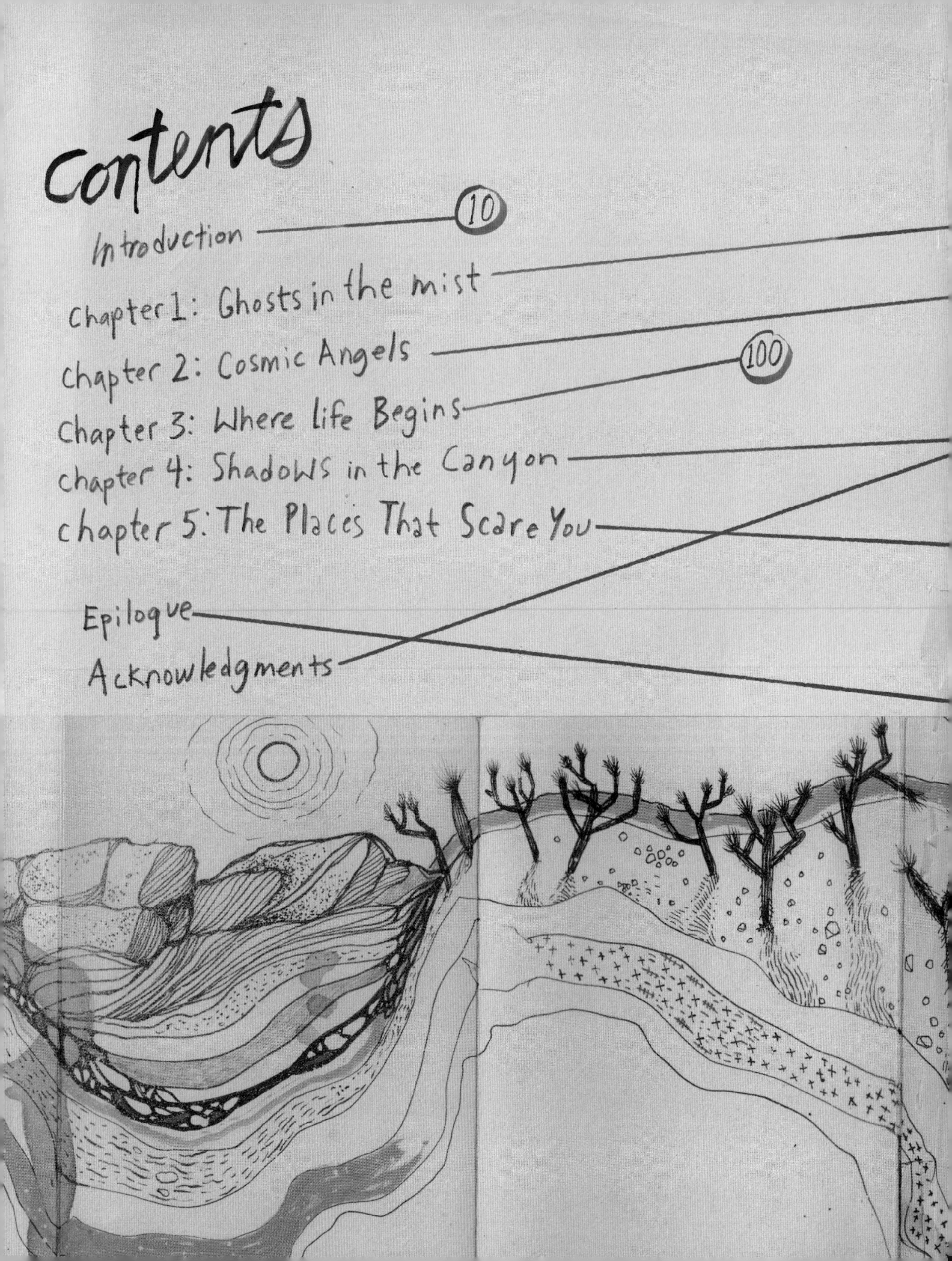

Contents

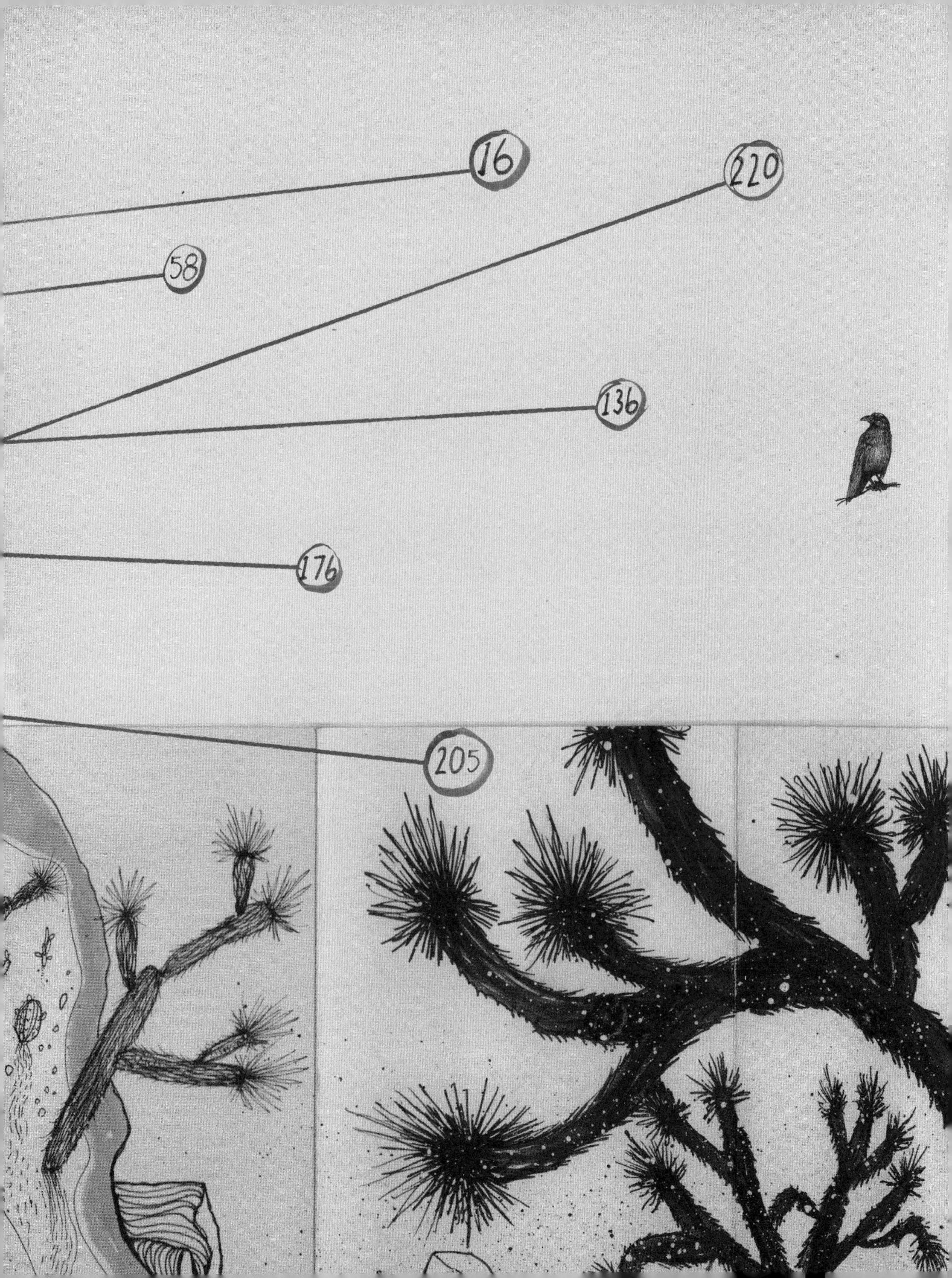
16
220
58
136
176
205

"IF YOU WANT TO LEARN ABOUT A TREE, GO TO THE TREE."

-Inspired by the Japanese poet Bashō

Introduction

"You'll NEVER make a living off this Shit."

My college drawing professor had an eloquent way with words. I attended a small midwestern university after landing a scholarship for art. Mr. Sample was a phenomenal draftsman, but if he had been a doctor, his patients would have said he had an unpleasant bedside manner. Prickly as a cactus, he would sit and analyze my work while dragging his bony finger across the page, occasionally smudging the drawings. (I knew he did this on purpose.) At the end of my third semester, when I went in for a one-on-one review, he looked through my season of work and grunted a few times, mumbling to himself. Then he looked up at me through his Coke-bottle glasses and dryly let me know how he felt.

He was probably right. My drawings were still childlike, with a naive cartoon quality, likely a direct result of my part-time job as a caricature artist. I didn't last much longer at university, but that was fine. I sent photocopies of my portfolio in manila envelopes to publishers and brands and received freelance job offers. I said goodbye to college after my sophomore year.

More than anything, though, I just wanted to go climbing.

In high school, I was introduced to rappelling on a roadcut along Interstate 70 near Columbia, Missouri. As semitrucks screamed by below, my classmate laced an eleven-millimeter rope around a large cedar and gently tiptoed over the edge of the cut. When I was halfway down the bluff, he accidentally knocked off a fist-sized rock from above that left a one-inch-long gash in my neck. Luckily, I didn't let go of the rope.

I arrived at the bottom with my T-shirt splattered in blood and yelled to him, "How about going up instead of down?" He just shrugged. Teaching rappelling in the summers at a Girl Scout camp seemed to be the extent of his knowledge of ropework, so if I wanted to learn to climb, I was apparently on my own. I headed to the library and found How to Rock Climb! by John Long and bought a climbing rope from the Campmor catalog. Five years later, I was drawing whimsical images to illustrate Long's articles in Rock and Ice magazine.

I guess Mr. Sample was wrong. To be fair, I was making "climbers' wages," which isn't really making a living, but at least I had found my niche. My first few publishing gigs helped me solidify my identity—they shaped how I saw the world and spent my time. I called myself a climbing artist and went on to illustrate hundreds of articles, books, and maps for various publications and brands.

I had found a way to survive as an artist, but over time, something felt missing. The bulk of what I created bordered on voyeuristic—made for someone else and about other people, random products, and things not connected to me. I was on the outside looking in.

After a few years of this, my younger brother Jonathan and I went on a week-long jaunt to the islands of Lofoten in northern Norway, inspired by a photo in a magazine. We were drawn by dramatic peaks diving into the Arctic Ocean, romantic villages, and promises of orca sightings. We didn't carry much and found ourselves sleeping in barns and at bus stops to save money. Although traveling light, I did bring a sketchbook, some pens and watercolor supplies, and a Polaroid camera to experiment with emulsion transfers onto paper.

Enamored with the swirling Nordic clouds and dramatic granite faces rising out of the sea, I returned home with drawings and writings that felt more like me—work that reflected a building storm inside my personal experiences and emotions, not someone else's. On the pages of that sketchbook was the sense of ownership I had been missing.

I started leaning into my own voice and telling my own stories.

Eight years after the journey to Norway, I produced my first graphic memoir, DRAWN: The Art of Ascent. In it, I shared four expedition-style climbing journeys, each in one of the four cardinal directions. The book and its accompanying film sent me all over the world, climbing new routes in far-flung destinations while pursuing a singular and soulful objective with close friends. I tried out new forms of media, creating work not just as an illustrator but also as an author, animator, and film director. With DRAWN, I stepped firmly beyond my identity as a climbing artist and into storyteller, no matter the medium.

As the DRAWN tour came to a close, I received an invitation to lend my drawing skills to an expedition that didn't include climbing at all. It was a water-based journey with an altruistic mission. "Do I do that?" I wondered. I felt equipped for the artistic challenge, but accepting the invitation would mean setting aside my climbing roots for a time. Yet there was a chance to tell a story that might really make a difference—what if I could do that?

In an increasingly tumultuous world, I had been questioning what the drawings in my sketchbooks could do to help amplify the causes I found moving, from human suffering to environmental threats.

I mean, they were just lines on paper.

Artistically I often feel like I'm running out of time to do my best work and that time is flying by. But "time flies" is not a thorough metaphor. Time doesn't fly. It explodes into fragments, then scatters about the floor under the furniture. It projectiles into the atmosphere and disappears in a breath. Time is a bullet exiting the chamber. But even that isn't explosive enough of an example. And I guess that's it—time doesn't fly or explode, it erupts. Saying yes to the invitation down the river was an opportunity to keep the lava burning.

Martin Luther King Jr. once offered some perspective for those of us who occasionally suffer from imposter syndrome: "Everybody can be great, because everybody can serve. You don't have to have a college degree to serve. You don't have to make your subject and your verb agree to serve. . . . You only need a heart full of grace."

Not many of us possess a heart FULL of grace (well, maybe my mom), but I wanted to see what was possible with my smidge of the stuff. I accepted the invitation for the river expedition and found it was a lot like my earliest days of climbing—I was following the lead of others doing great and graceful work.

In the years following that river trip, I've had more opportunities to travel with inspiring folks and document their stories in an ever-growing stack of sketchbooks. These people definitely have more than a smidge of grace. They have taken on objectives larger than themselves. And yet none of them started out that way. In fact, each one of them started small, like the seed of a giant sequoia tree.

The giant sequoia is the largest tree species on the planet and the largest organism by mass. The biggest ones we see today first sprouted from the earth two to three thousand years ago, when Roman emperors ruled most of what would become Europe, people still believed the world was flat, and my college professor Mr. Sample was probably grading cave drawings. On the other side of the world, in a quiet forest once inhabited by dinosaurs, sequoias started their long climb to becoming towering ancients.

They were small saplings with immense potential to grow into something greater, just like the people I've met, whose stories you'll find here. Matt Redd's "something greater" utilizes photography, cattle, and science. Craig Childs crafts the written word. Bernadette Demientieff translates legend and history. James Q Martin and Céline Cousteau create documentary films. Prem Kunwar provides education and disaster relief. They are just a few of the creative individuals who are trying to make a difference in the world today.

I didn't meet these "sequoias" by accident. I either went looking for them or manifested a connection through the work I was pursuing. Then I had the immense honor to follow them with a pocket of pens and a desire to grow. When we loosen our grip on who we know ourselves to be, we make room for the possibility of becoming something more. Following the individuals in this book into wild places helped me do that.

-Jeremy

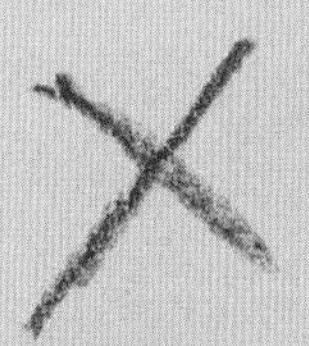

CHAPTER 1
GHOSTS IN THE MIST

"In the Marubo cosmos, the forest canopy is the horizon par excellence, the conflation between nai 'sky' and mai 'earth.' Within the tama shavaya, 'arboreal dwelling' or 'clearing,' stands out the shono tree (. . . samaúma in Brazil). . . . This top stratum of the forest is the intermediary cosmic layer between the earth of the living and the rejuvenating skies of yové-transformed, peeled-off yora-bodies. Beyond . . . the sun follows its paths . . . , the overarching, supreme dwelling "sky of clouds" koĩ nai shavaya enveloping the whole native cosmos. The land of the living lies down below all that."

— Guilherme Werlang da Fonseca Costa Couto
EMERGING PEOPLES: MARUBO MYTH-CHANTS

Maybe I would be able to sleep tomorrow. I was curled up in the fetal position around an empty fifty-five-gallon gas drum on top of our thirty-foot-long fiberglass boat. There was a dense mist floating on the Amazon river, making navigation impossible. Perhaps we were intersecting with the Death Path—a misty purgatory where, according to a Marubo shaman, deceased bodies are stuck while their spirits ascend up into the samaúma, the tallest tree of the Amazon.

Every few seconds, pink boto dolphins surfaced for air, delivering a dramatic exhale from their blowholes. PSSSHHH! From the shoreline, a large black jacaré (alligator family) slithered in and out of the reeds for its nocturnal hunting—snatching delicious piranhas or turtles and pulverizing their bones in its powerful jaws. I triple-wrapped my wrist into the webbing on the gas drum in case I dozed off too soundly and plunged into the water, but sleep was just another mirage in the fog for now.

Hours ago our crew of nine ran out of fuel and started paddling, first as a joke and then from necessity to avoid getting caught atop massive, submerged palm trees. Beyond the fog, a sliver of moonlight gave a ghostly glow—enough to see just inches beyond the bow but not much else. Beneath the surface, limbs hid like mines waiting to tear a hole in our vessel. I was wrapped in a tarp, with my satchel as a pillow and all my clothes layered on. At my feet at the other end of the drum was a soggy and quiet Céline, sleeping soundly despite the chaotic sounds surrounding us.

BORN of WATER

Céline Cousteau is the granddaughter of Jacques Cousteau, the famous French undersea explorer who is often credited with introducing the world to the mysteries of the ocean. His television programs, books, and movies featured expeditions aboard his equally famous research ship, the Calypso. From jet-sized humpback whales to elusive octopuses to streaming schools of speckled fish, the underwater world he and his team revealed inspired generations.

Céline was only nine years old in 1982 when she and her father, Cousteau's oldest son, Jean-Michel, joined her grandfather and grandmother Simone (an accomplished researcher and diver in her own right) on a two-year exploration of the Amazon, the world's largest and longest river. The effort was dedicated to sharing the ecology, wildlife, and peoples of the Amazon basin with a wider audience. Céline's own childhood memories from the journey, however, are fragmented, like a blurry, click-through slideshow in her mind.

"I remember taking the Zodiac boat to visit a man harvesting bananas on the shore line"

she told me. "He invited us to his home in the jungle, and his wife fed us a meal over a fire. For the longest time I had no idea what we ate, but it's funny how the mind works—decades later I took a bite of manioc (yuca), and it sent me into a time-warp portal of memories. That's what it was! I remember the barefooted children playing games with a stick, and I was too shy to join. I remember a scientist catching piranha and them splashing in the shallow water beneath my feet in the boat. The people and the ecosystem anchored my relationship with the Amazon. I didn't know it then, but a door had been opened to my future self." Thirty years later, Céline was back in the jungle, soaking wet and curled up opposite me on our boat's aluminum roof. Perhaps it felt like a homecoming to her—the waters of the world pulsing through her veins as they have through her whole family. But this time she was not a little girl playing chess on the deck of the Calypso. She was the captain, leading her own expedition born of her own vision.

Céline wears as many hats as her legacy dictates—explorer and film director, of course, but more than anything, she is a storyteller.

Here in the Amazon, the story was NOTHING NEW—Western civilization's encroachment on Indigenous peoples. For centuries, outsiders have come looking to indoctrinate the remote tribes found throughout the greater Amazon Basin, or to extract what resources they can from the rivers and forests. Either way, they have left a wake of disease and discontent. The Brazilian government seems to put their collective fingers in their ears, deaf to the needs of their own people's health and cultural necessities.

CÉLINE COUSTEAU

COLOMBIA
GUYANA
RORAIMA
TABATINGA
RIO BLANCO
MARUBO
RIO NEGRO
MANAUS
RIO AMAZONAS
BRAZIL
PERU
N
BOLIVIA
PARAGUAY
CHILE
ARGENTINA

At a conference in Rio de Janeiro, Marubo and Matis tribal members approached Céline and asked to tell their story in the hopes that their voices would finally be heard. A year later, she arrived with a film crew, a photographer, an anthropologist, and me—an artist. Our purpose was to navigate the Javari region of northwest Brazil and amplify the voices of those most in need. Céline plotted a two-week journey down the Rio Novo and Javari tributaries of the Amazon river to visit three specific tribes: the Matis, Marubo, and Kaminari.

I was invited by Céline to "do what you do, but without a climbing harness. I want you to observe, find a connection, and draw." She followed this up with "On second thought, if you wanted to bring a crossbow and rig a line up into the jungle canopy for an aerial perspective, you might bring some climbing gear just in case." But I didn't. After a decade of climbing expeditions, the idea of traveling light was immensely compelling to me: to leave the gizmos and ropes and gear behind and just focus on supporting Céline's objective with hand-drawn imagery crafted from the experience.

We survived the foggy night, and our fuel resupply boat arrived to keep us going south. As the miles blurred by, I quietly drew and reflected on the previous three days in the village of Boa Vista with the Marubo people. Our final night we were invited to the maloca (community house) for a parting gathering I can only describe in terms of my upbringing in Missouri—it was a barn party. We joined the families for a ceremony consisting of circle dancing, chanting, and smoking copious amounts of rapé (pronounced "ha-pay"), the endemic tobacco plant that is mixed with other forest ingredients. All the local Marubo were in full ceremonial attire, including beaded headdresses, full-body painting, ornamental jewelry, and piercings. The women dressed Céline and our anthropologist, Barbara Arisi, to match and brought them into the pulsating circle to join the dance. The film crew, photographer Michael Clark, and I lingered mostly in the shadows, watching the line of figures as they moved around the maloca. A shaman shared a brown mash liquid among the men, made from the vine of the ayahuasca and served in a gourd. It tasted like wheatgrass but sent my brain spinning.

The flames of a fire in the center of the room amplified the sinewy shadows of the dancers (or perhaps it was the shaman's juice), and the space filled with a profound, vibrating energy. Sitting on an old stump, rocking back and forth to the music for a few moments, I forgot why we were there. Before us was resilience in motion of a people who have endured outside threats since first invaders arrived from Europe in the sixteenth century. Despite modern medication being readily available not too far beyond their remote jungle location, over 80 percent of people in the Javari suffer from all variations of hepatitis. From every angle, their traditional ways of living are threatened; they have to deal with Indigenous health risks, biodiversity threats such as pollution and poaching, and illegal extractive activities in the jungle.

It was Beto Marubo, a Javari Indigenous man, who first approached Céline at the conference and asked her to help share his people's story. Beto is a coordinating member of the Union of Indigenous Peoples of the Javari Valley and a previous staffer of the federally appointed National Indigenous People Foundation. But more important, Beto is deeply committed to the health and survival of everyone in the region. He knows the threats intimately and wants the needs of the community to be seen beyond the jungle and beyond Brazil in the hopes that such publicity will help drive change.

If Beto provided our invitation, Barbara was the key to helping us unlock the voices of the jungle. Barbara is an Italo-Brazilian professor of anthropology who has lived among the Matis, the neighboring tribe of the Marubo. She can speak the Javari dialects, translate, and advise us on local customs. She was welcoming and lovely, always smiling and joking, and had a bottomless wealth of knowledge on all things Amazon. Barbara was considered family in many of the villages we visited and was greeted warmly. She also taught us about the dangers of any unplanned stops along the shoreline, as they could result in encounters with "uncontacted" tribes. In Brazil, the government uses "uncontacted" or "unknown" to describe any group who chooses to maintain a remote and traditional lifestyle with no relationship with the larger national society.

We were even instructed not to make eye contact with a tribe named the Korubo, should we accidentally meet in the area. They have been known to blow darts or bludgeon invaders without question, and they are referred to locally as "the Head Bashers," using what they call a borduna, a type of weapon made of wood. I peed off the back of the boat obediently and didn't ask for any stopovers.

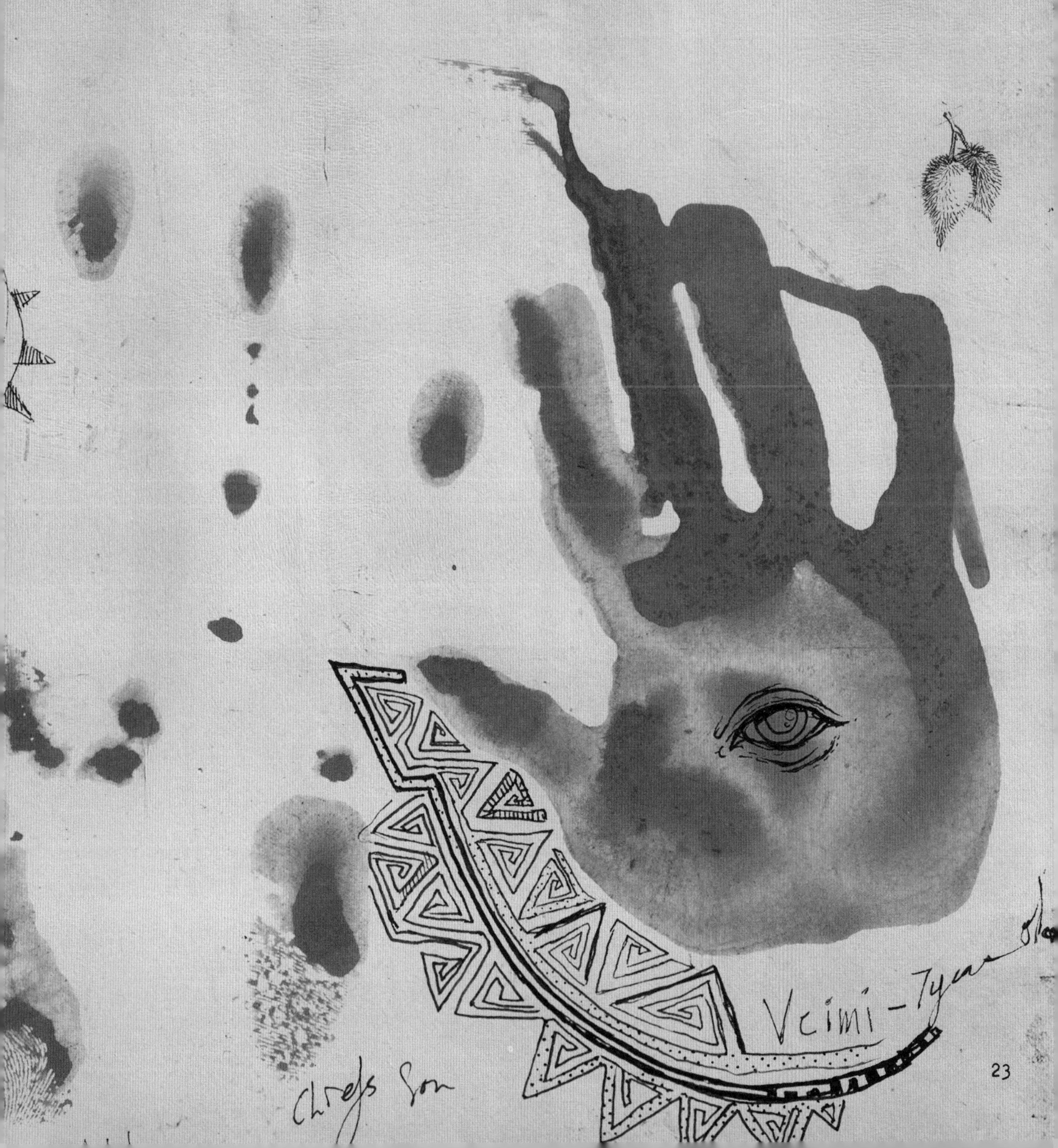

When we arrived at each village, we were ushered from the boat to the main house, where Barbara explained to the village leaders what we were there to do. A discussion would follow as to what that community wanted the world to know—although each village consistently asked for government assistance in providing medicines and health care. Céline always listened intently, with pen and paper in hand, committed to sharing their story accurately based on their own words. In these initial meetings, she set aside her film director hat and was strictly a listener.

Our camera crew kept their gear packed until a plan for the day was in place and permissions given. I found a nook nearby to stay out of their way and just observe. When I pulled out my pen to draw, I was often **flanked** by children curious as to my activities. I started simple, with the trees framing the village. Inevitably, one of the children would lean against me and place an elbow on my shoulder. **WITHOUT** Barbara to translate, I allowed for the language of art to be our communication. If it felt right, I pulled out blank books and my supplies for the children to draw alongside me. Part of my role during the expedition had become village icebreaker, as my time with the kids usually led to laughter, and eventually the children began sharing their hunting skills, or introduced their pet monkeys.

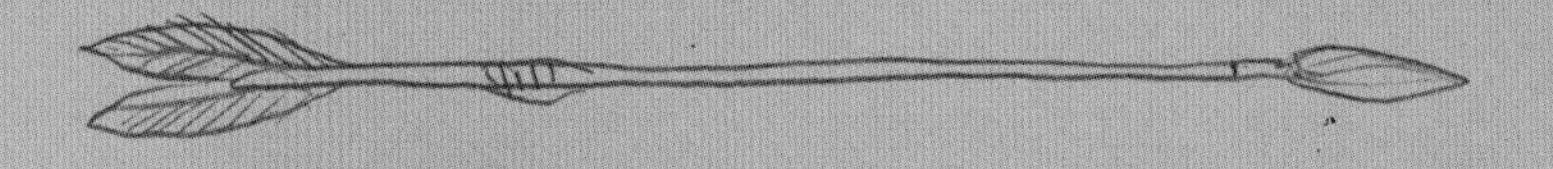

When the sun hit, I hid under the canopy of Brazil nut and pupunha palm trees. I went for quiet walks through the village just to listen to birds, laughter, and the chittering conversations of women in huts. I joined a pickup game of soccer and made friends with a young spider monkey who spent an hour on my shoulder and then pooped on my shirt in gratitude.

At one point I was offered a shot of rapé tobacco, which was administered by staring into the eyes of the one serving as a long, tubular bird bone was inserted into your nostril. The server then violently blew the entire load into your brain. Céline went first, and with red watering eyes and a green ooze coming out of her nose, she encouraged me to go next. I decided to try to follow Céline's lead and remain stoic through the experience. The Marubo man serving me clenched his fist in response and said, "Forte" (strong). It burned like an inhaled shot of whiskey through a straw, and I had a buzz for the next hour. When no one was looking, I finally gagged.

Inside, Each stall represents one family group. They have cooking utensils, banana supply, the day's meat kill, 2-4 Hammocks and large poles for constant fire. The embers stay 24 hours a day to keep bugs away, then get stoked for cooking.

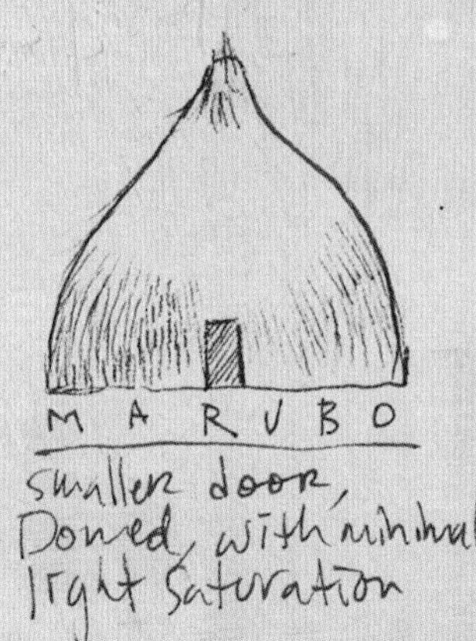

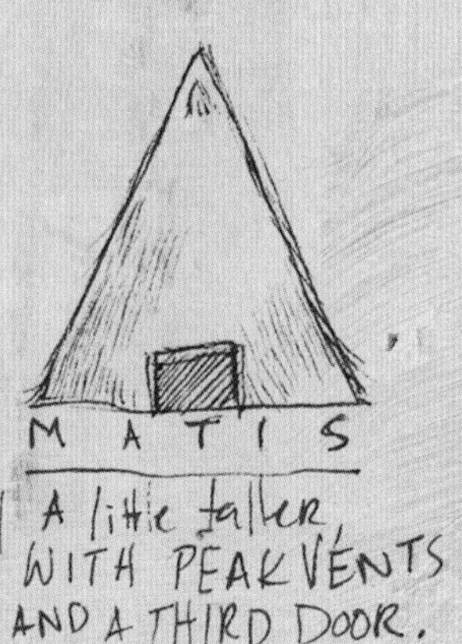

Wildlife played the role of companion and sustenance. Large tapirs roamed the villages like the family dog. Children were given a monkey (pygmy marmoset) when they were young to care for; the animals spent their days gripped to the top of their owner's head and screamed at me if I approached too quickly. A hunter returned with a dead toucan hit by his arrow, and I tried to hide my gasp. NOT TOUCAN SAM! Children returned from riverside hunting outings with handfuls of large bullfrogs and turtles. After a meal of turtle soup, I went out back of the hut we were in and saw piles of monkey skeletons presumably from previous dinners.

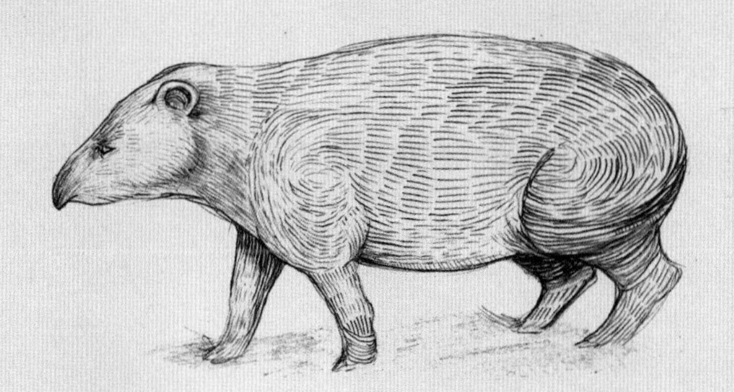

Common Name: TAPIR IN PORTUGUESE: ANTA OR ESPAÑOL SACHA BACCA
OR AS I LIKE TO CALL IT: JUNGLE BACON

The Javari region has the largest population of Indigenous peoples living in voluntary isolation in the world, and we weren't the only ones swooping in to see how we could help. In fact, when Jacques and company arrived on the scene in Brazil to document the river all those years ago, just across the border in Peru were my maternal grandparents, Larry and Eunice Bryant, working as Christian missionaries in Lima.

My grandparents worked throughout South America from 1945 to 1985; their four decades of international travel plus their conservative missionary perspective greatly influenced me when I was young. Growing up, I felt, Why travel to a developing country if you aren't there to help? Perhaps Céline felt the same way. Rock-climbing expeditions eventually broke me of that mindset, although I certainly explored ways to integrate positive impact into my personal objectives. Joining Céline on this trip, however, I finally felt like I had a real purpose that relied specifically on my particular skills—drawing and the ability to be productive while uncomfortable. In fact, one of Céline's first questions was, "Are you okay sleeping in zero-star accommodations?" I assured her I was.

"DON'T CALL US JAGUARS"

In Some Sense, I have no business telling you about the MATIS people I encountered. I am not an anthropologist, and I was BARELY a visitor. Instead, I hope to CONVEY a sense of WONDER and KNOWLEDGE about something you might know already. After all, I was just another dumb white man trying to help.

Remote peoples have been in these Riverways long before the modern recording of their presence. ARCHAEOLOGISTS say people have lived in the AMAZON for more than eleven thousand years. I was there for under two weeks. Everything I learned came from Reading Before and After and FROM info given from BARBARA, BETO, or Céline.

Otherwise, my knowledge comes only from my brief moments with boots on the ground. But remember, I was there to observe, find a connection, and draw.

And what I observed is the Matis do not want to be called "the Jaguar People," as they have come to be known by journalists. The Matis are master hunters and warriors, though, with a striking appearance, so one can understand how the term has lingered. Identity deaths don't come easy. Historically, their ancestors utilized face painting, acai palm tree piercings, large shell earrings, and whisker-like tattoos to mimic the elusive black creatures lurking under the canopy of the forest. Modern Matis still feature these distinct markings and are the most fascinating people I have been fortunate to meet in remote situations. They were not overly excitable—they greeted us and wanted to get right to business telling their story.

All beings have a driving instinct to exist and should be granted the opportunity to do so with support from other beings if needed. I think that is one of the things that makes us human. Reportedly there are over one hundred uncontacted tribal groups in the Amazon, and even if they are not threatened by diseases from outsiders, they still have to deal with poachers, illegal mining, drug traffickers, intertribal warring, and the obvious threats of living in the jungle with little to no protection. It's not beyond me that we were a boat full of white people and cameras telling the story of the impacts of the same thing. Sound familiar? How much has changed since both Céline and my grandparents came to the region to "help" decades ago? Were our grandparents here with the best interests of the Indigenous in heart? Or are the "uncontacted" better just left alone to thrive or die on their own cosmic timeline? It raises the classic existential question, if a tree falls in the woods and no one is there, does it make a sound?

In his book The Unconquered, journalist Scott Wallace observed, "Indigenous groups living in isolation are isolated because they choose to be. It's not for complete lack of contact, but precisely because previous experiences of contact with the outside world proved so negative." Wallace based these statements on the weeks he spent following Brazilian explorer and activist Sydney Possuelo, a foremost expert on isolated tribes in Brazil who has dedicated his life to contacting and protecting them. For his efforts, Possuelo has earned dozens of awards and accolades, and likely enemies as well. He remains adamant that "Indigenous people have never faced a worst moment in Brazilian history than the one they are now facing." The culmination of years and years of devastation to the Amazon and its rainforests—devastation now heightened by the impacts of climate change, along with all the relentless human encroachments—have pushed the Matis and other tribes to the edge.

I can't help but wonder if the worst threat to ancient culture is its inevitable crash into the modern one. And yes, now that includes me. If modern society does not romanticize or report on the uncontacted with pretty pictures and films, what is the global loss?

Is holding onto and protecting traditional ways of living important for our collective human future? The questions buzzed in my ear like the mosquitoes over my hammock as I watched Céline diligently do what she had been asked to do—document the threats in the Javari and tell the world.

But . . . everywhere you step in the jungle lies a threat.

The large things were certainly obvious: the black caiman that lingered dangerously close to our boats on the shoreline, the jaguars we never saw but assumed were ever present, the schools of piranha, the anaconda, the numerous venomous snakes. Then there were the Head Bashers and, even deeper into the jungle, the Flecheiros (the Arrow People), who have no contact at all with the modern world and shoot any guests with a swarm of arrows.

The list goes on and on, but the most notable discomfort for us was far smaller—even smaller than the swarms of mosquitoes and the horribly painful biting ants. The locals called them "piums," perhaps a distant cousin of what I know as no-see-ums, except these insects seemed descended from fire-breathing dragons. The invisible demons burrowed into our skin constantly, rendering our appendages unrecognizable, as if we were made of Mr. Potato Head parts and had replaced our hands, legs, and feet with swollen, oozing plastic versions. The women in the village nibbled the insidious bites off the body of their partners when they settled down for the evening, while we just itched and sprayed our useless bottles of ointments. (It's worth noting that this may have been the first time I wanted someone to nibble something out of my feet.)

One afternoon we were treated to a tattooing ritual in the big house. Tattoos were a central axis of Matis identity—the lines and colors signified their community affiliation, life experiences, and transition from adolescence. The tattoos were applied using acai palm tree thorns dipped in a black paste made of a mixture of a resin and a compound obtained by burning various native plants. The parallel-line motif was called muxá, signifying knowledge and strength passed from one generation to another. They called this legacy strength their xó. It was striking, beautiful, and singular. Whereas I usually am drawn to more natural elements in my journals, I couldn't deny the irresistible allure to create portraits of the beautiful faces of the Matis.

a Marubo Woman in Boa Vista Village
Rio Novo, Javari, Brazil

With two weeks behind us on the river, we had grown accustomed to a rhythm: rise at dawn, eat quickly, and get to work. The camera crew was in a constant state of preparation and archiving. They utilized solar panels to charge batteries and accommodate nightly uploads of footage. Each morning, I followed the instructions gleaned from the great artist Barron Storey: "Wake up. Draw. Draw yourself awake." Although quite present with the experience, I was often immersed in my black Moleskine journal. If I was not drawing, it was at all times in a small satchel over my shoulder and under my arm, zipped up in a waterproof sleeve. I carried ten pens of various sizes, a few quarter-ounce watercolor tubes, a brush, and two pencils. Included in the satchel were a monocular, headphones, extra batteries, and a small pocketknife.

I have made a career that is sustained by drawing or painting images for commercial, editorial, or publication use. Honestly, I don't know how to do anything else. To wake up and draw images that had the audacious potential to alleviate human and ecological suffering felt like more than drawing myself awake. I was here to draw myself alive. This was my purpose, and Céline saw it. Once the sun hit, however, my sole purpose was to stay cool.

Céline never stopped interviewing. She talked to anyone who wanted their voice heard. She said she was going to take a break but never did. In the heat of the day, covered in oozing pium bites, no matter how much someone wanted to say, she was listening, asking questions, and taking notes. Maybe she was hiding any sign of fatigue, but the rest of us were absolutely exhausted.

One morning we woke to the sounds of wailing. I couldn't tell what I was hearing. A bird? We were informed it was a mother mourning the death of her child. If we felt any curiosity of our purpose there, whatever the reason, perhaps the cries of a mother reinforced Céline's heart and intent.

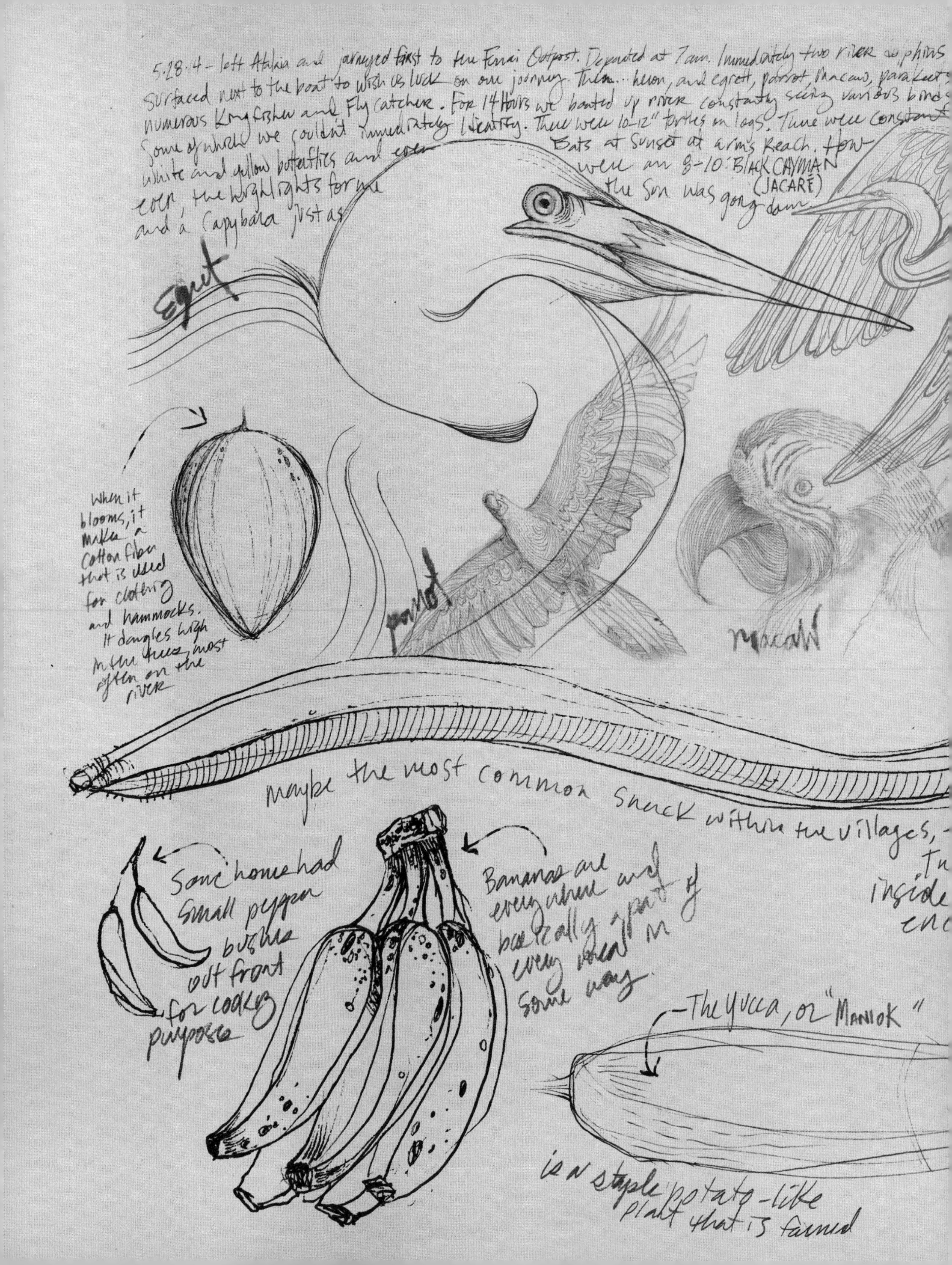
5.28.14 - left Atalaia and journeyed first to the Fenai Outpost. Departed at 7am. Immediately two river dolphins
surfaced next to the boat to wish us luck on our journey. Then... heron, and egrett, parrot, macaw, parakeet,
numerous Kingfisher and Fly catcher. For 14 Hours we boated up river constantly seeing various birds
some of which we couldn't immediately identify. There were 10-12" turtles on logs. There were constant
white and yellow butterflies and even
even, the highlights for me
and a Capybara just as
Bats at sunset at arm's reach. How
were an 8-10' BLACK CAYMAN
(JACARÉ)
the sun was going down.
Egret
When it blooms, it makes a cotton fiber that is used for clothing and hammocks. It dangles high in the trees, most often on the river
parrot
macaw
maybe the most common snack within the villages,
Some homes had small pepper bushes out front for cooking purposes
Bananas are everywhere and basically a part of every meal in some way
The yucca, or "Maniok"
is a staple potato-like plant that is farmed

Piranha
Kingfisher
2
TARTARUGA MARINHA
DOIS REAIS
River dolphin
"BOTO"
dangle from trees everywhere.
three feet long with an
ng seeds with a tasty pulp
y.
unpermanent yet bold,
these are used for painting
faces. I saw red, green and
yellow.
tart, and
sour fruit
used for
snacking
and cooking
delicious fruit with
a center of suspended
seeds.

As the sun started to go down at the end of each day, our group of seven non-locals would migrate to the river, red-faced, sweaty, and itching. We had been told it was safe to swim and bathe at the shoreline, and the large groups of children who spent time at the water daily had made us confident, but we were still on high alert. We tried to make bath-time fast and efficient. I had a small bottle of biodegradable soap and a quick-dry hand towel. We didn't chat too much when bathing because despite the water feeling absolutely glorious on our cooked and bitten bodies, we didn't want to be in what felt like a precarious position for too long. We turned off our headlamps and quickly got it over with.

After a week of this, we operated fairly seamlessly as a unit, except one night in the village of TAWAYA, one of us was missing.

8:4

"WHERE IS BARBARA?"

someone asked.

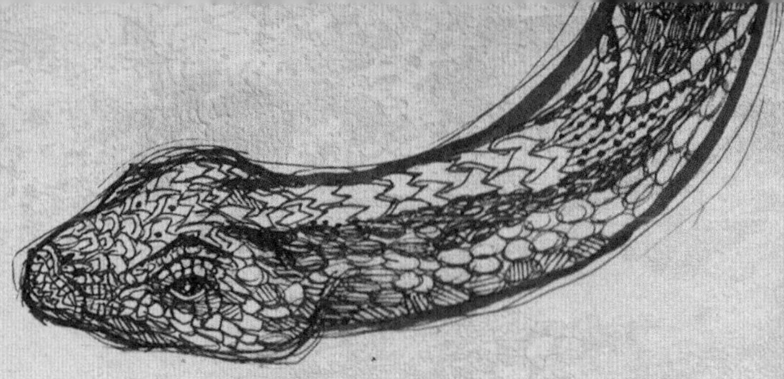

Practically immediately, she announced herself in a most violent way from the dark grassy hillside above. A spine-tingling scream rang in the air, and we scanned the darkness but saw no flashlight. Then a follow-up that sprang everyone into action: "Snake! Snake! It bit me!" In a flurry of splashing water, we all exited the river. Our cameraman, Capkin, arrived to her first, and sure enough, she had two bleeding fang marks on her lower leg. I swung my headlamp to the tall grass in hopes of identifying the species. I was barefoot and soaking wet. I took two steps and quickly thought better of it.

Barbara started screaming for the village nurse Felipe. Barbara's adopted Matis brother Tëpi picked her up and hurriedly carried her back to the village, a couple hundred meters away. The entire village came to life in the dark. The men donned rubber boots and ran into the tall grass with machetes to look for the snake. After the standard one-hour wait, Felipe administered a generic antivenin that the village had in stock. The snake was not located, but based on the elders' input, Barbara had been bitten by a surucucu, or pit viper, one of the most venomous snakes in Brazil.

The threat was potentially fatal.

TBT AM
VR|46

Felipe quickly called via UHF radio to a neighboring village, asking for help. There was no way to know how much of a load of venom had been deposited into Barbara's leg. Did she have an hour? A day? Could she survive the night? An eight-horsepower boat left the Indigenous village an hour upstream of us with more medication, as our boat simultaneously took off into the night to meet them halfway. Barbara was hung in a hammock, sweating profusely yet trying to stay calm. Her leg began to swell and blacken, a sure sign of decay setting in quickly. Felipe and Céline set up the satellite phone with the intent to initiate a rescue. Eventually, they were able to reach the Special Secretariat for Indigenous Health office, and Céline hurriedly requested a helicopter. After a fifteen-minute debate, it was confirmed: because Barbara was not Indigenous and was visiting on Indigenous land, a rescue would not be approved. Our team and the locals were on our own. Felipe held back one dose of the antivenin just in case someone else was also bit before we returned. Barbara had meant to pick up our own supply before leaving the city but ran out of time.

At sunrise, we moved Barbara into the other boat and quickly began the fourteen-hour journey reversing all the way we had come. She was stable, but her leg was black and her skin flush, with constant beads of sweat across her forehead. In true Barbara style, she still joked and giggled, but then would wince under the severe pain. A hammock was strung between two vertical supports in the boat, and she hung across us all in a sweaty group cuddle. I began drawing and writing down the experience so I wouldn't forget the details. We made one stop for gas at a waterside outpost and hustled in to take bathroom breaks. Upon entering the small building, Céline tugged at my shirt and whispered, "Korubo. Do not look." Of course I glanced. So did she. There they were—a Korubo family curled up on the floor, with their signature half-bowl haircut and wearing nothing more than what can be made from the forest floor. We were told they too had come out of the jungle looking for medical help.

After the brutally long journey, we arrived at the town of Atalaia do Norte and were informed there was no doctor in the city, so it was back into the boat. We finally arrived at Benjamin Constant, another twenty-five kilometers north on the Javari river, where a public hospital took Barbara in and administered multiple vials of antivenin. They said she likely wouldn't have lived without it. I couldn't help but feel like we were just another boat of white folks coming to fix previous white folks' impact and we still were a threat. What if someone died because we hadn't come prepared?

The irony hung over us like the canopy of the Amazon forest. We came to tell the story of the lack of medical support available to the Indigenous of the Javari, but we consumed most of the stored antivenin for two villages equaling at least one hundred people. How long until they could get a replenishment? What if someone was bitten the following day or in the time we had been gone? Although it was Barbara's plan to bring antivenin, I kicked myself, and all of us, for being so foolish to not have had our own. Céline paid for the replacement, but who knows how long it would take to arrive.

Our boat puttered back into the modern world of Manaus, a city preparing for an upcoming World Cup soccer match. There were banners and signs all around town, with festive drinks being served in the bars and a reported $3.6 billion spent by Brazil to update the stadiums and infrastructure. We slowly motored through the colorful favelas (slums) where trash poured from the neighborhoods into the Amazon. Here the houses were built on stilts using whatever material could be found or harvested from the city.

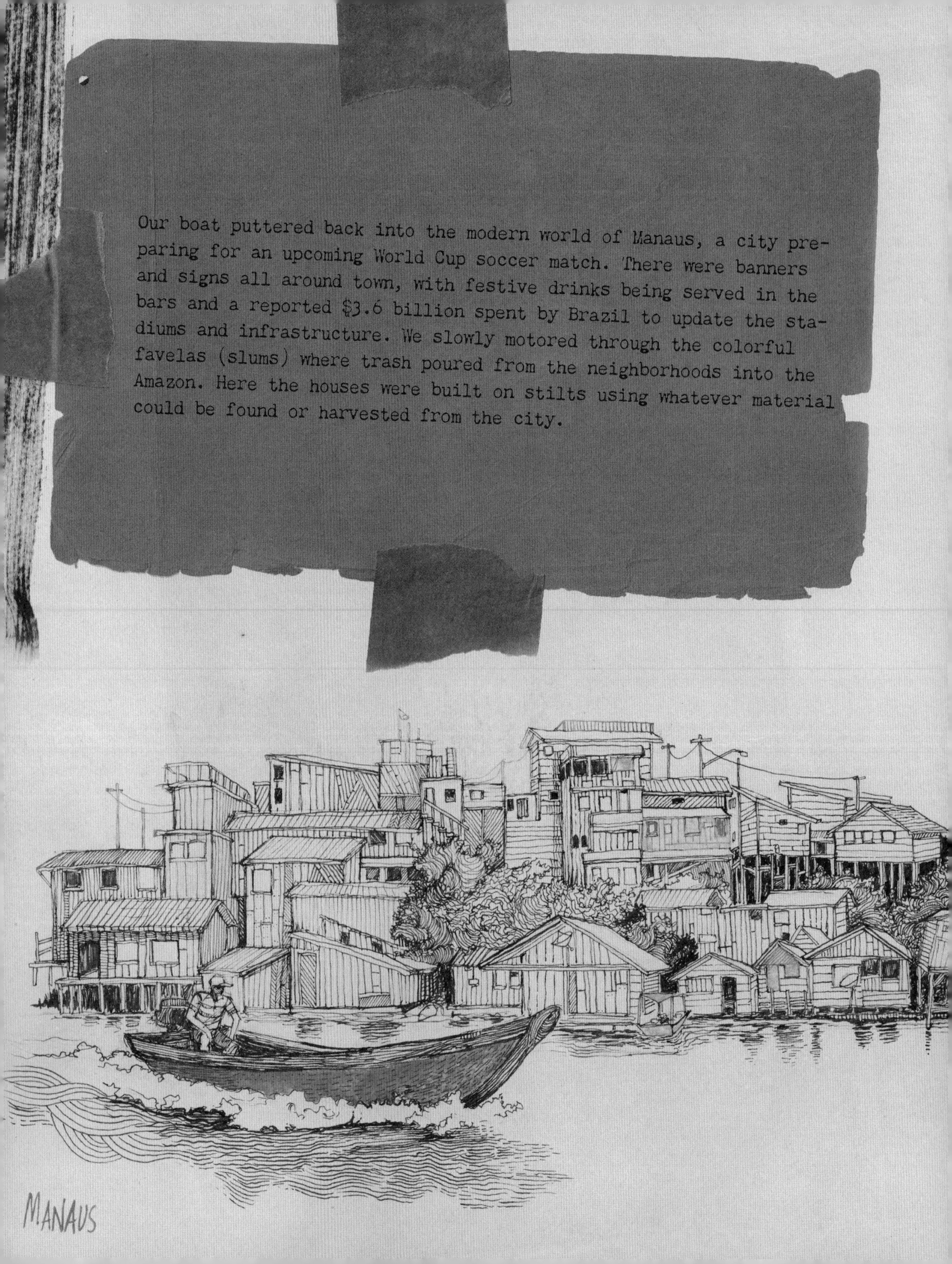

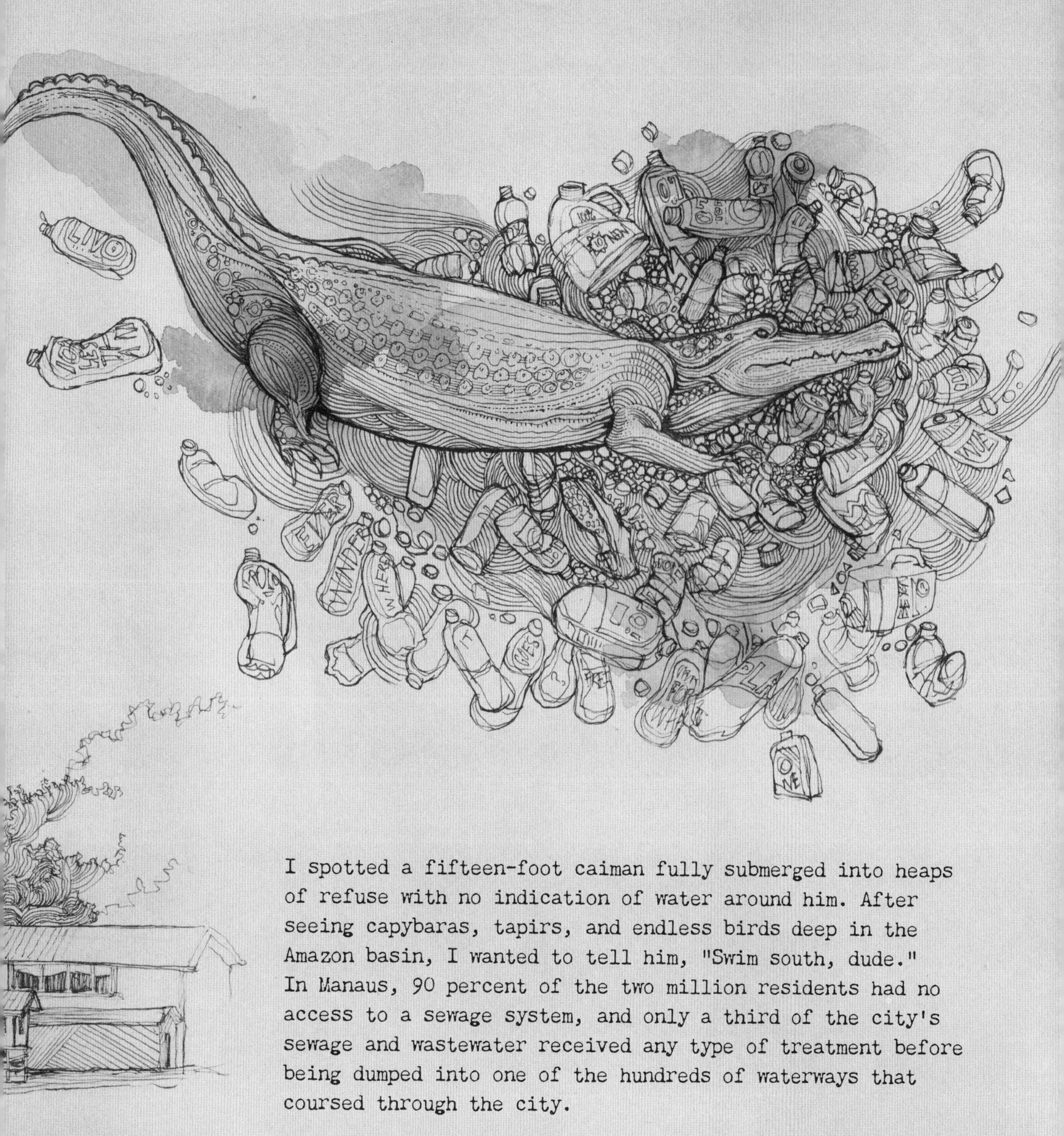

I spotted a fifteen-foot caiman fully submerged into heaps of refuse with no indication of water around him. After seeing capybaras, tapirs, and endless birds deep in the Amazon basin, I wanted to tell him, "Swim south, dude." In Manaus, 90 percent of the two million residents had no access to a sewage system, and only a third of the city's sewage and wastewater received any type of treatment before being dumped into one of the hundreds of waterways that coursed through the city.

Céline described everything we were seeing in ecological and social context as she continued to scribble notes and DOCUMENT thoughts into her recorder. We had become siblings on the journey. I made fun of her being sweaty and covered in bug bites, trying to save the world, while sitting in our aluminum boat and simultaneously reading the latest copy of ELLE magazine with her legs CROSSED.

She embraced the joking and LAUGHS—entirely comfortable with her one foot in the bougie WORLD of fame and the other in the business of giving a damn. With the camera team, she began discussing "next time" and "after the WORLD CUP." She was scheming on going farther downriver to uncover more of the truth, which of course, she did.

MUNICÍPIO: Benjamin Constant, Atalaia do Norte, São Paulo de Olivença e Jutaí		8.519.800 Ha	2.068 km
		ESCALA: 1 : 550.000	DATA: Maio de 98
UF: Amazonas	AER: Atalaia do Norte	PROCESSO Nº: 1074/80	BASE CARTOGRÁFICA: SB.18-X-B e D, SB.18-Z-A, B, C e D SB.19-V-A, B, C e D SB.19-Y-A, B e C

TÉCNICO RESPONSÁVEL PELA DEFINIÇÃO DOS LIMITES	TÉCNICO RESPONSÁVEL PELA IDENTIFICAÇÃO DOS LIMITES	CONFERE: CHEFE DO DED	PORTARIA :
WALTER ALVES COUTINHO JUNIOR ANTROPÓLOGO - DAF	SEBASTIÃO CARLOS BAPTISTA ENG. AGRIMENSOR CREA SP 77.417/D	MANOEL FRANCISCO COLOMBO ENG. AGRIMENSOR CREA SP 64889/D	174/95 e 158/96

Our team disbanded and traveled forward in time to hot showers, cold drinks, and streaming movies, while Barbara spent the next month in the hospital recovering from the venom coursing through her veins. After seven years of dedication and conviction, Céline released her independent documentary titled Tribes on the Edge with the approval of Javari leadership.

In the film, the translated voices of those in the Javari share their perspectives:

"We have seen the white world on a television, and there are many, good things. But here in the forest, it is much better."

"The POLICY of the current GOVERNMENT is development and economic growth at all COSTS. There are large political interests in these agencies, and that hinders all the WORK. IF you look, government investment has increased by millions, but in the VILLAGES, there is nothing to SHOW FOR it."

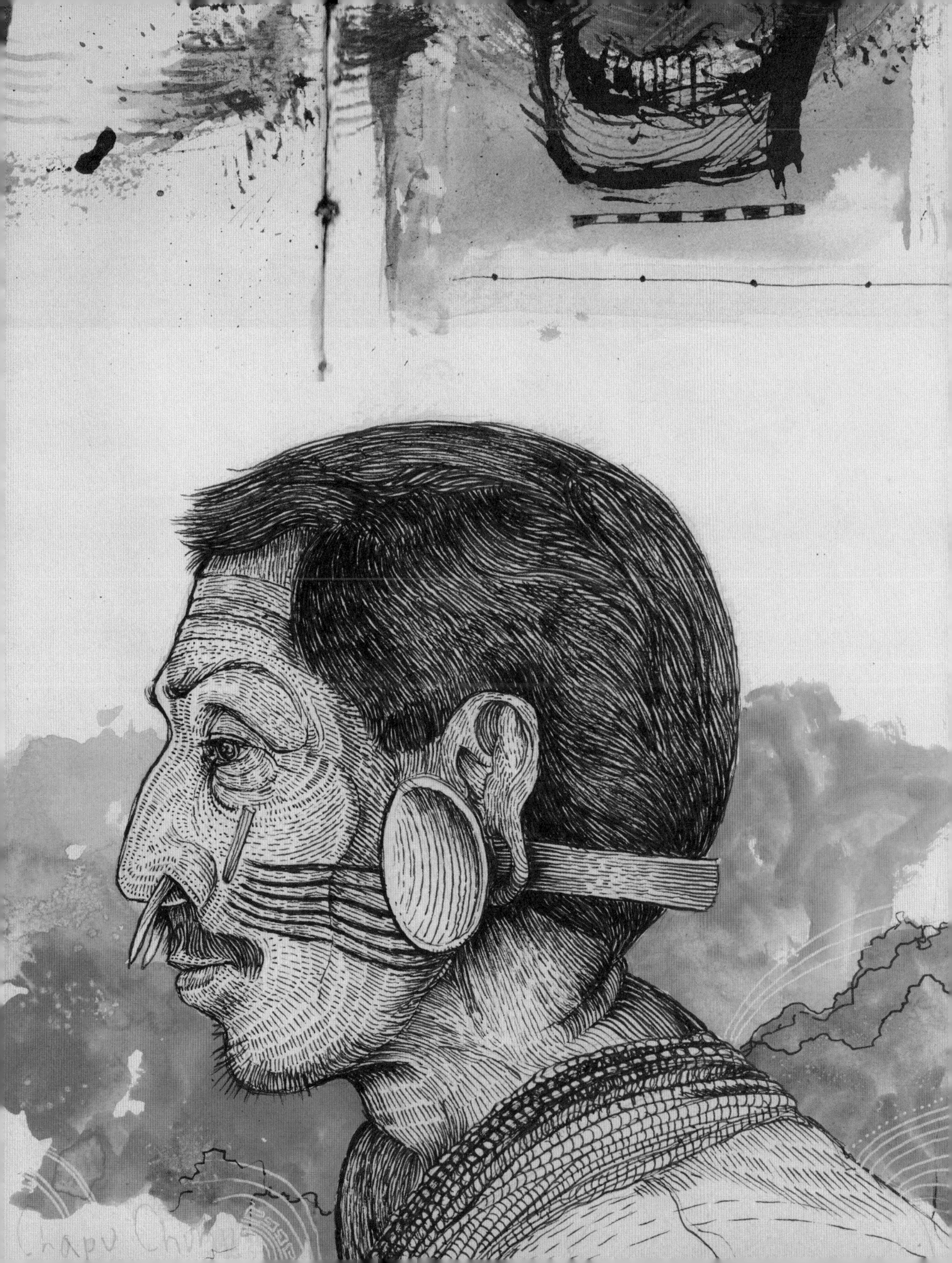
Chapu

"We are the main preservers for these woods, so we want the government to recognize that we are important for its survival."

"Our history is very long—a people that was almost extinct, ended. My parents were massacred by whites. As I am the chief, I'm struggling to see the future of our own children taking our people forward."

"We have enough. Bananas, native fruits, manioc. WE DO NOT NEED the THINGS from the CITY. The things we grow are good for us and our COMMUNITY."

"I have met several outside people who want to help us, but the BRAZILIAN government won't let them enter the region. I'm 28 years old and all of the relatives close to me are dead."

Céline's film became a catalyst for the Javari Project, a coalition of partners to implement change in the region, offering schools and delivery of antivenin to the most remote villages of the jungle.

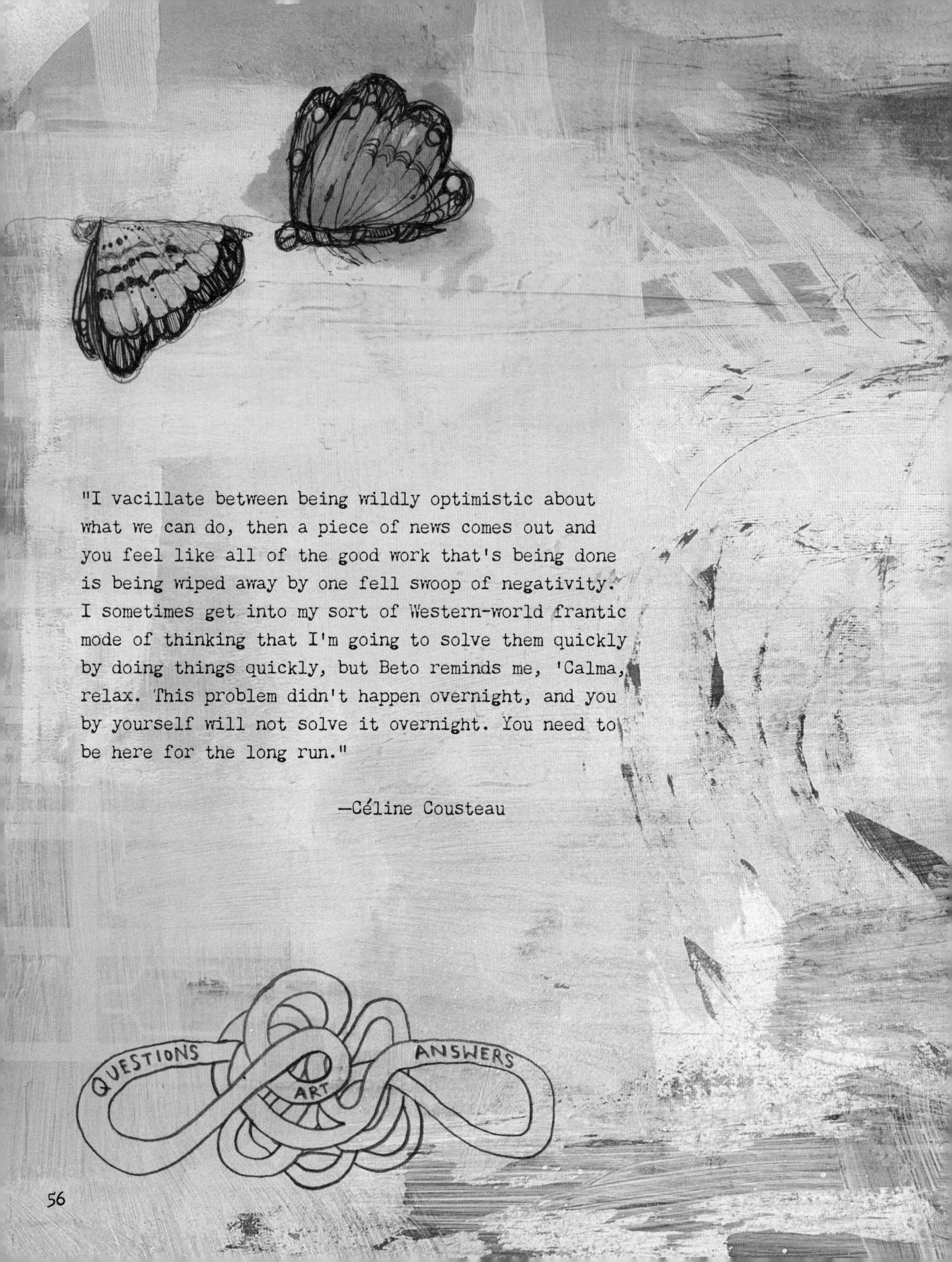

"I vacillate between being wildly optimistic about what we can do, then a piece of news comes out and you feel like all of the good work that's being done is being wiped away by one fell swoop of negativity. I sometimes get into my sort of Western-world frantic mode of thinking that I'm going to solve them quickly by doing things quickly, but Beto reminds me, 'Calma, relax. This problem didn't happen overnight, and you by yourself will not solve it overnight. You need to be here for the long run."

—Céline Cousteau

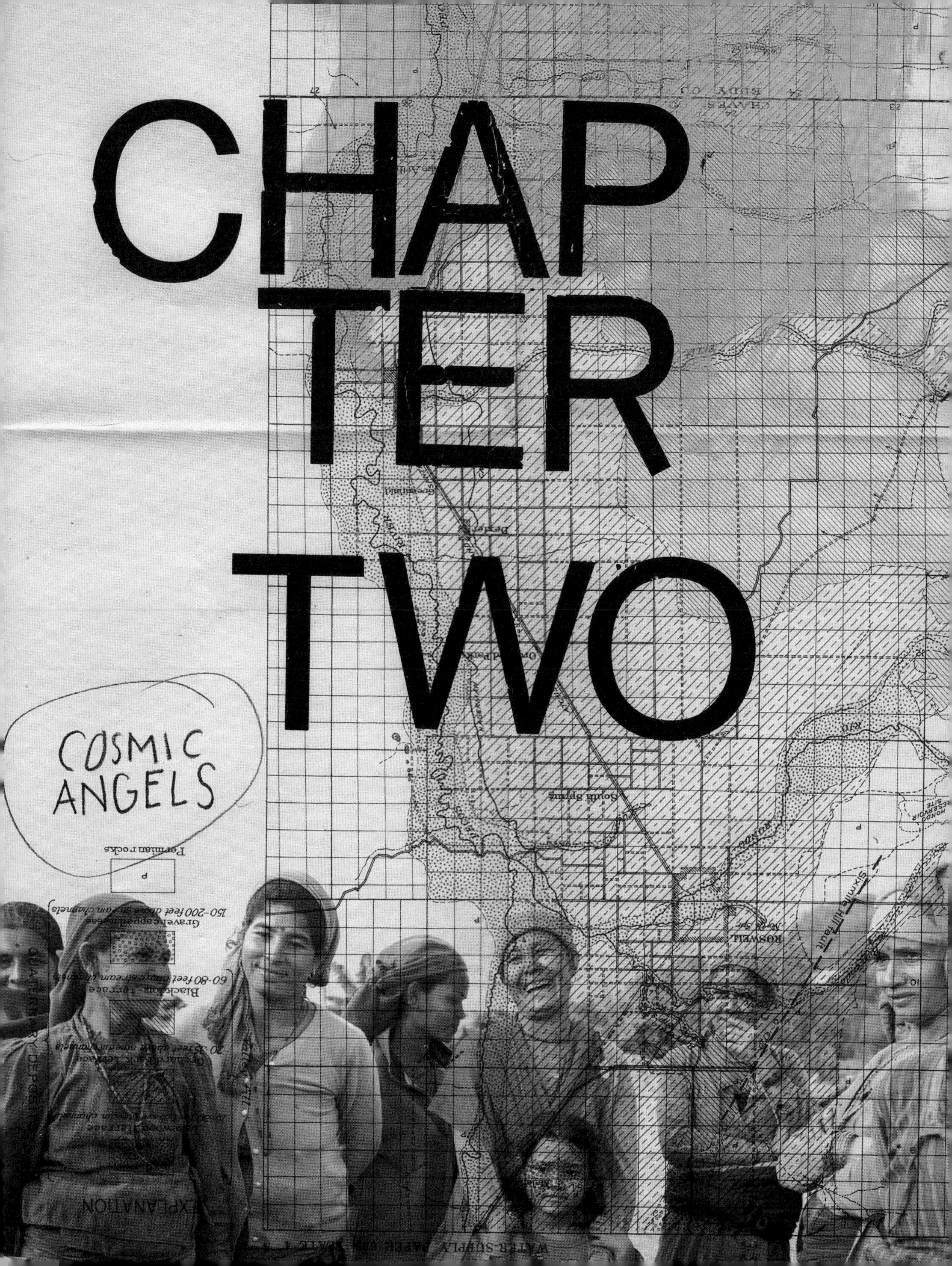
CHAP
TER
TWO
COSMIC
ANGELS

"No, I'm sorry, Jeremy-Dai, but you cannot climb that mountain." Prem Kunwar patted my back in humored consolation as I sat and drew perhaps one of the most aesthetically appealing peaks I had ever seen, featuring massive walls of granite interspersed with snowfields that led to a cat-eared double summit. Its Nepali name is Machapuchare, which translates to "Fishtail," inspired by the iconic split summit, and it is considered a sacred place to be left untouched. According to local lore, Machapuchare is the home of Lord Shiva, protector of the region, who some legends say still resides among the summit boulders. The windswept plumes of snow that can be seen coming off the summit are thought to be the smoke of Shiva's divine incense.

At seven thousand meters, it's certainly not the tallest peak in the Annapurna range, but it is the most pristine—entirely absent of peak baggers' trash and decaying corpses. Whether or not the mountain was an option for me on this trip, I was not there to climb. I was there to observe, find a connection, and draw. Céline Cousteau had cracked the door open, and now I was exploring a new way of traveling with my time capsules (a.k.a. my sketchbooks) as an ad hoc journalist of sorts. Again, the harness and climbing gear stayed at home.

Far beneath Machapuchare, at a much lower elevation and on the banks of Phewa Lake, Pokhara is an attractive tourist destination with far less pollution and poverty than that of the immense Kathmandu, a six-hour bus ride away. The streets are alive in Pokhara with a distinct bohemian heartbeat. Music from the world pulses from coffee shops and boutiques. Dreadlocks and flip-flops are popular, and incense wafts in and out of doorways. Prayer flags and beaded gift items sway in the breeze.

In the middle of Phewa Lake is a small island that houses the most important place of Hindi worship in the region: the Tal Barahi Temple. The two-story temple is made of stone and accessible only by boat.

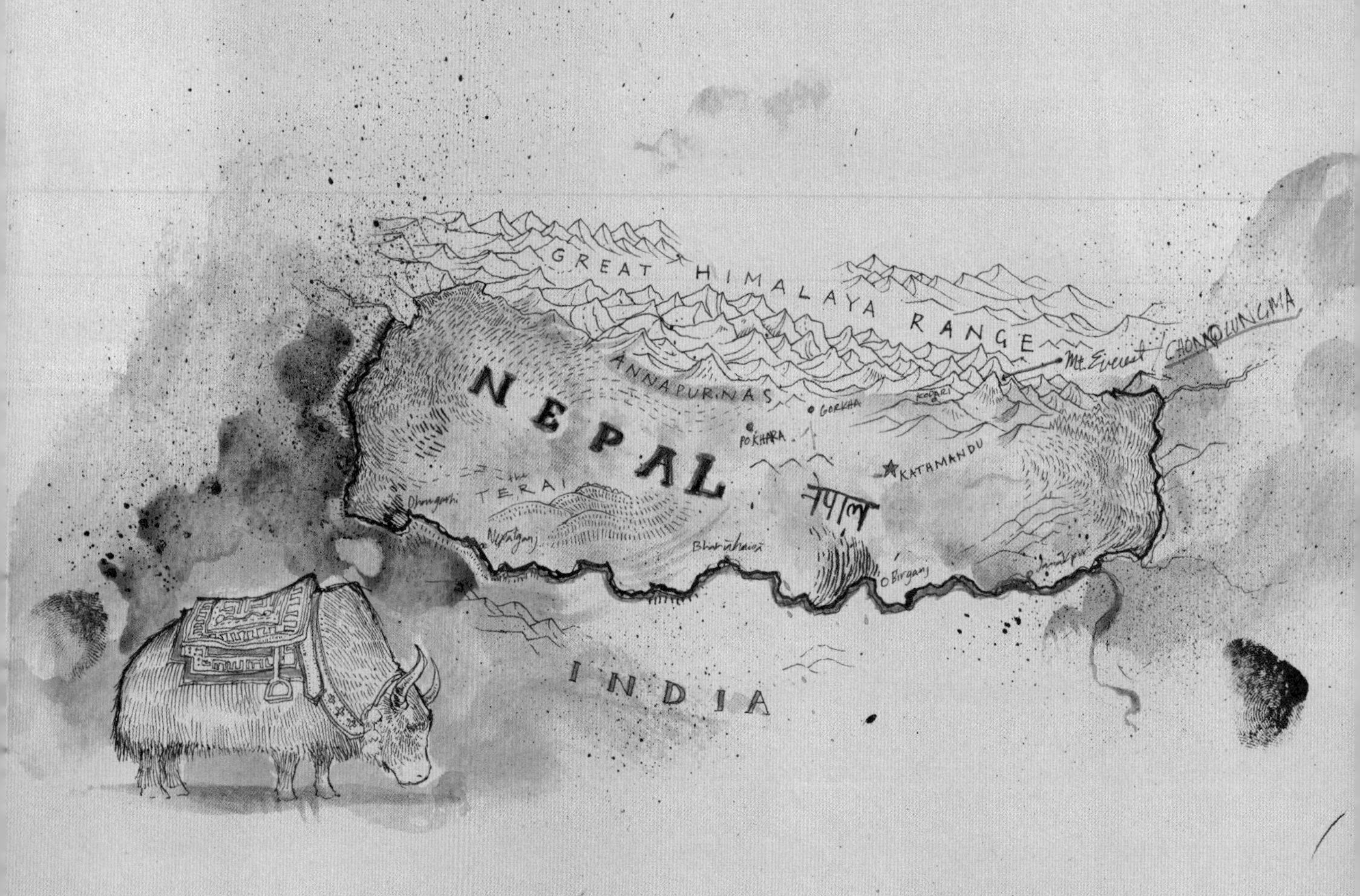

Beyond Pokhara in the hills of western Nepal, Prem was born in the remote village of Arnakot, home to only a few hundred people. He remembers social service being a central part of his upbringing and recalls his family planting banyan trees for shade and to create community meeting places. Arnakot was a four-day walk to reach anything considered a road and didn't have a school. Plenty of kids would see this as winning the no-school lottery, but not Prem, who was thirsty to know what the world offered. He and his siblings walked an hour to attend the nearest elementary school and two hours once they reached middle school. When it came time for high school, Prem and two of his friends were determined not to let distance keep them from their education, so they walked barefoot 168 kilometers to Pokhara. In the bustling tourist city, Prem was exposed to the outside world and fell madly in love with learning.

After advancing his education and working diligently at a local restaurant in Pokhara, Prem began sending a portion of his earnings to the teachers who taught in his first village school. Over the years, his efforts, along with others' support, transformed the community, which now boasts a well-equipped library, a solar-powered computer lab, disaster-resistant school buildings, and a team of excellent teachers catering to over three hundred students. After seven years working as a translator for the UN Human Rights Office, Prem opened the Cosmic Brontosaurus Language School in Pokhara with his Australian friend Greg McGrath. The school is ideal for both travelers wanting to learn the local dialect and local young children wanting to learn English.

PREM KUNWAR

In the spring of 2015, some of Prem's students were in town for paragliding, one of the popular tourist sports in the region. Jamie and Isabella Messenger were lifers—pilots who lived to fly and enjoyed teaching others. With consistent conditions, beautiful launch points, and an alluring atmosphere, the Annapurna range is perfect for flying. The couple met in Pokhara while Isabella was involved in a parahawking project (training rehabilitated birds of prey to fly with paragliders). She had heard about "this Jamie guy who had just won the Canadian National Championships" and looked forward to the chance to learn from him. They joined each other on a tandem flight and talked all the way to the ground. "We chatted the whole time and I knew I liked him, but he had to pass one test. I gave him my favorite novel, War and Peace, to read, to make sure he was not just a pretty face and amazing pilot. I would see him every day riding on the roof of the tandem jeep with big torn sections of the book, as it was too chunky to fit the whole tome in the back of the harness. Done deal."

The Messengers were eventually married and started a nonprofit with Prem and others called KarmaFlights with a goal of providing safe homes and educational support for local children in need. Little did they know how well positioned they would be on April 25, 2015.

Outside the office doors, clients were waiting to go paragliding. In a nearby apartment, Jamie was taking a quick shower when his world began to shake. He ran into the street covered in soap suds with just a towel around his waist. Isabella said the violent earthquake was like being "an ant on the back of an angry elephant." Within minutes, the shake was over, and everyone in their vicinity seemed okay. If it wasn't for their friends immediately texting reports from international media, they would've had no idea that the destruction was far worse in the higher altitudes than in the lowlands of Pokhara.

When the dust settled across the region, over nine thousand people were pronounced dead and over twenty-three thousand injured in NEPAL and surrounding areas. It was the most deadly natural disaster in Nepal's history. Then the rains that followed the quake were unrelenting. In the mountains, entire villages huddled under plastic tarps, unable to return to their homes under the threat of Aftershocks and trapped by massive mudslides that blocked entry or exit.

The Gorkha district, a mountainous zone between Pokhara and Kathmandu, was hit hard. Homes in this area were built by piling local flat stones and logs into rectangle homes. There was no Home Depot down the road. They used what the land provided. The quake demolished these homes and community buildings immediately, many with people tragically inside.

Within forty-eight hours of the tremor, the team of paragliders and Cosmic Brontosaurus teachers assembled supplies and began making their way toward the epicenter; others jumped on dirt bikes and four-by-fours to join them. The mountains looked like a war zone, with homes splintering into piles of rubble and children aimlessly wandering the dirt roads looking for family.

The chaos was overwhelming, but they were diligent in their conviction that they could help.

With Prem leading the charge, the KarmaFlights team, along with other pilot guides, drove through the night, intent on doing whatever they could. They located a doctor and a couple of nurses to join them. Despite having tickets to return to the states, Jamie and Isabella had no question in their minds: we stay and do what we can. They canceled their flights and determined to stay in the region indefinitely. A month became a season became a year. The team established a medical triage care center and a scholarship program for orphaned children, and they delivered tons of food and water to remote locations. They also began school rebuilding projects utilizing earthquake-proof sandbag construction.

As I watched the images from Nepal via online media from the other side of the world, I felt an unavoidable urge to help, but beyond a meager donation and social media post, I felt helpless. I felt art could help . . . I just wasn't sure yet how. I knew I'd be going, but I didn't know when would be most useful. Nepal had received over $4 billion dollars in relief fundraising from around the world. It would take years, perhaps decades, for the country to be completely rebuilt.

POKHARA
SPEED
CONTROL

Nepal is the ultimate destination for both the serious and armchair mountaineer. Any truly passionate climber is lured to Nepal eventually, whether for an attempt at Everest or another eight-thousand-meter peak or to make the practically mandatory pilgrimage to base camp. For mountaineers, Nepal is what Yosemite is for rock climbers like me.

I have never had the mountaineering bug for the big peaks, or what I like to call "Major Hiking Achievements." My drive is simpler and far warmer—I love ROCK climbing: the less wet and white stuff, the better. Desert towers, yes. Granite big walls, yes. Steep cragging, yes. Big mounds of snow and ice glued to the sides of a big hill? Pretty, but no. I've lost too many friends to those unpredictable environments. That said, Nepal beckoned me. It always has.

Six months after the disaster, I shared my still-lingering intent for a visit to the epicenter with my friend Nick Greece. His eyes lit up as he shifted to the edge of his seat. "You HAVE to do this, and you NEED me to go with you."

After the quake, KarmaFlights partnered with the Cloudbase Foundation, a US-based organization whose mission is to empower hang glider and paragliding pilots to be effective agents of change within communities where they fly. With the logistical and financial support from Cloudbase, KarmaFlights expanded its scope to encompass emergency disaster relief and scholarships for orphaned kids in the surrounding districts.

A Bronx native and world-class paraglider, Nick had been one of the accompanying founders of KarmaFlights back in 2011 and now was the conduit connecting me to Prem and the Messengers. Whereas I intended to simply follow my nose through Nepal, Nick had a vision for how to make the journey purposeful and multiply our impact.

We quickly hatched a plan that included my longtime traveling partner James Q Martin, a filmmaker who could document and amplify the experience with hopes of generating further support for those most victimized by the quake. Known most commonly as "Q," he was more than willing to share stories of need in the world. Prem met us upon arrival and was ecstatic to see us. His warmth and genuineness were clearly authentic—I was immediately drawn to him. We headed to Cosmic Brontosaurus to meet his team and plan for the coming weeks.

I felt as if I had stepped from the quiet banks of a lazy river directly into a rapid current of joy and wholesome motivation. The enthusiasm for our being there was robust, and Prem quickly shared ideas of how we could help. Practically all of the schools in the area were reduced to rubble, and many of the makeshift rebuilds had been constructed from donated corrugated steel. I didn't want to say it, but they looked like prison camps. It was explained to us that many children in the Gorkha were now scared to even go to school for fear of another tremor collapsing the schoolhouses onto them.

It was decided we would bring paint and beautify the schools, but first, Nick and the Messengers wanted to show Q and me why Pokhara was so special to them.

We packed our backpacks with overnight gear and two sets of tandem paragliders. We hired two porters to help carry the load and started the two-day trek to the base of the sacred Machapuchare at thirteen thousand feet. Q, Nick, Jamie, Isabella, and I left Pokhara, heading northwest into the rhododendron forest with full packs and excited hearts. I'd never paraglided before, and despite trusting my guides, I felt a bit uneasy.

We were there to help, not risk anyone's safety. But the trek was beautiful. We passed water buffalo, crossed creeks, and circumvented fields of rice. We waved at farmers and shared a reciprocal "Namaste" with our hands clasped in prayer to all we came in contact with. We could see the remains of old buildings and the humble beginnings of new ones.

We slept for the NIGHT in a trail-side bunker built for trekkers headed into the higher ANNAPURNAS. We made dinner at a communal picnic table and found space to CREATE. Q filmed clouds through the trees, NICK double-checked equipment, and I drew until I was too cold to be productive. The following day, we broke through tree-line with a full view of the utterly stunning Annapurnas.

MOUNTAINTOPS surrounded us in a dizzying glow as we traipsed ever upward through swaying gold grasses. A tall stick was mounted in the ground with primary-colored prayer flags whipping along the LENgth of it.

As the sun set on our second night, Nick and I looked out over the valley we had come in through as it disappeared in a blanket of clouds. I pulled out my pen and paper to draw but instead just sat silently to take it all in for fear of forgetting the moment. Ravens soared above us, and I watched as a pair playfully ended the day much like us—absorbed in the vast space. I asked Nick what it was like to fly. He described a weightlessness and purity of freedom. He talked about updrafts and downdrafts and thermals. Then the undeniable Bronx in him kicked in: "It's fucking amazing. You're going to love it." That's what I was afraid of—loving it too much. I had enough hobbies that concerned my mother already. He described the technical aspects of checking altitude and barometric pressure, how the variometer works, and how a landing zone is a known unknown. The logistics reminded me a lot of climbing—just follow what you know, and everything will be okay. It's a balance of experience, creative decision making, and detachment.

Nick saw my unopened sketchbook and nodded toward it, saying,

"I guess it's a lot like drawing—the freedom of the blank page."

"I can see that. What if we did both at the same time?"

"What, fly and draw?"

"Yeah, is that crazy?"

The last glimmer of cerulean light melted away into abstract shapes on the frozen shoulders of the mountains. The temperatures started to drop, and we melted into our sleeping bags. Nick and I decided that in the morning, with my sketchbook, two pens, and our GoPro cameras at the ready, we would fly tandem and I would attempt to draw during our flight. I fell asleep to the sound of Q puking his brains out as a not-so-gentle reminder that we had gained three thousand meters of elevation in our trek to get to this point.

At sunrise, Nick's variometer chirped information in morse code, and I paced nervously. I was glad he understood what it was telling him because I sure didn't. Q and his camera would be tandem flying with Jamie, I would be flying with Nick, and Isabella would be flying solo. I zipped my sketchbook into my jacket front and put a pen in a thigh pocket. I considered finding a string to tie to it, but the winds were good, and we had to go now. It was time to fly. Rising in a dramatic arc to our left was Machapuchare. I hoped Lord Shiva didn't mind our visit to his sacred airspace.

Nick did some final checks, and I strapped into my harness, then connected myself to his front side like a giant baby carrier. Nick yelled, "GO GO GO" in my ear. We waddle-ran to the edge of a grassy cliff until there was no more and let go of the earth. The parachute grabbed hold of the air, and we were aloft, quickly rising into the sky. We had launched from thirteen thousand feet and were already one thousand feet higher. Nick adjusted the straps as I kept remarking eloquently,

"WOW, WOW, WOW." He giggled at my excitement.

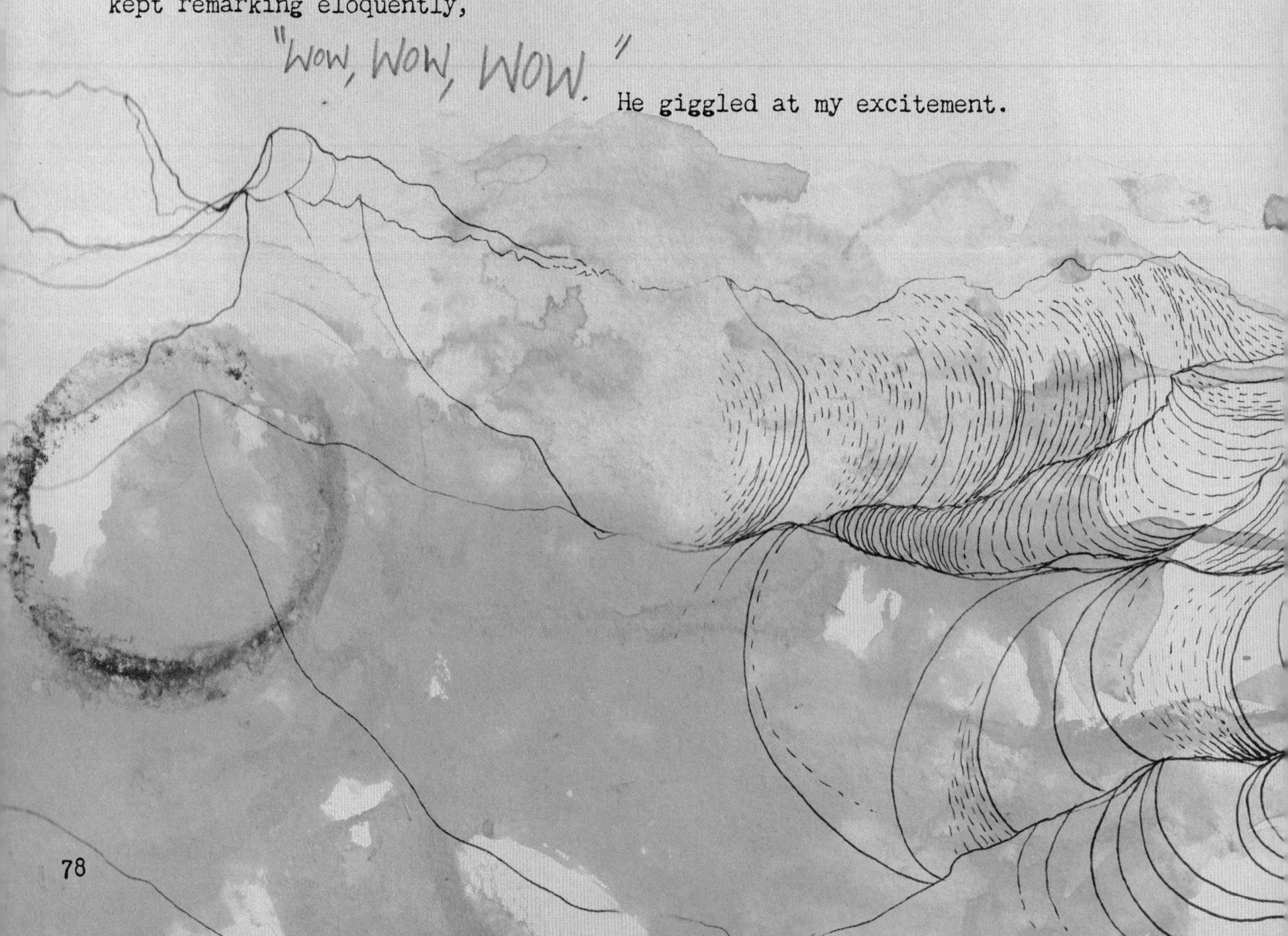

We stabilized into a more vertical position, and I could see the rest of the team flying near us.

Nick said, "Well, I bet we have forty-five minutes. ARE you READY to draw?"

"I'll try." I pulled my sketchbook out of my jacket and opened it on my lap. Then I dug into my thigh pocket for my pen and immediately felt nauseous. I let Nick know. "Bleh, not yet," I said. Four thousand feet of air was under my toes, and I knew it was a crapshoot whether I could pull this off.

I took some deep, cleansing breaths and lowered my eyelids halfway to impress upon my brain a sense of calm. A technique that I employ during stressful climbing is to "relax the skin on your face." This tells the rest of your body what to do. Nick encouraged me, saying, "If you are going to do this, you probably better get started." I opened my eyes and pulled out my pen, immediately entering "the zone" with the familiarity of my most common tool of expression.

All noise ceased except maybe the imagined strike of a Nepali singing bowl.

My heart rate slowed.
A montage of sights and memories flickered through my brain . . .

My very first season of learning to rock climb with my late brother.

My children's first laughs.

I sensed the sound and grit of a pencil dragging across the surface of paper as heavy as a cinder block.

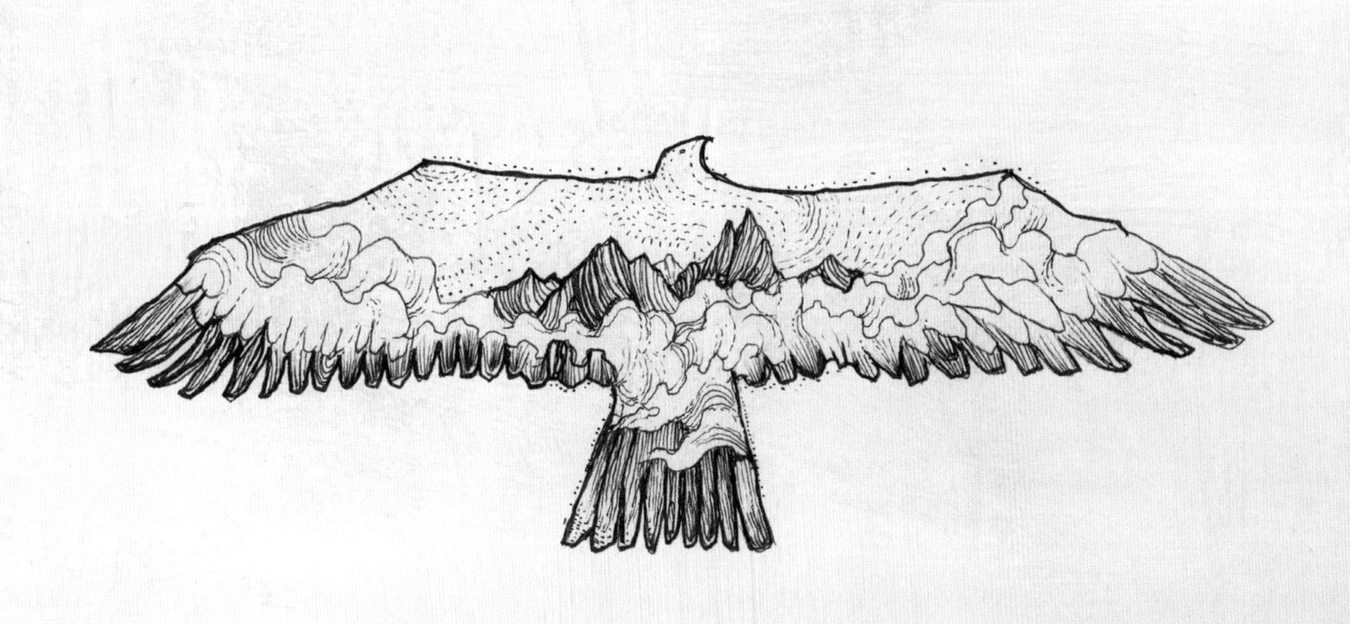

Any stresses or doubts I had, even the fear of something going wrong on this flight, all became mute for the moment. I was absolutely present and had become a falcon. Feathers sprang forth from the skin on my hands. I blinked my eyes, and they reopened with vision eight times stronger than my own. I flexed my back and rolled my shoulders forward as if I had wings. I began drawing the simple outline of an in-flight bird and then filled in the negative shape of the body with the scene unfolding in front of me. I called this "thing in a thing." Nick quietly piloted us and let me enjoy the moment, seeing that once I began drawing, I relaxed. It was a stunt that took no rehearsal and no real skill other than maintaining the simple curiosity about whether we could do it.

I finished my drawing and stuffed it away. We spiraled back to earth and landed in a farmer's field. After packing up our chute, we walked to the closest road and hitchhiked back to town. There we met Prem and headed to our first school, just above Pokhara in the foot-hills. We had paint and brushes and quickly got to work with a group of rambunctious and excited kids.

The schoolmaster said, "How about some smiling faces on the school? The kids are scared to go, and maybe that will help." We spent the next couple of hours painting two children in bold yellow and blue together. I was disappointed with the image. It felt like a cartoon and not the kind of work I was capable of. The headmaster was grateful, the kids had fun, and we made a cold gray building colorful, so who cares what I felt like. I wasn't there for an ego boost but to be of service. I took note and detached myself from the image and attached myself to the experience. Observe, find a connection, and draw.

It was dark out, and everyone was packed into a white pickup truck like sardines—Prem, Q, Nick, Isabella, Jamie, and Nick. I was curled up in the truck bed. Our driver, "Yam the Man," had helped on many journeys into the Gorkha for Cosmic Brontosaurus, and was one of the drivers that joined the Messengers when they were first responders. Almost immediately, we got stuck at a creek crossing, and all of our headlamps came out as we helped shore up the tires spinning in gravel. It was raining lightly, and we checked our maps under an umbrella. Before too long we were back in the truck and miles into the Gorkha district, the tourist comforts and niceties of Pokhara behind us. I looked at the people I was traveling with and was humbled and grateful to be there. Nick was right—I needed him.

We arrived in the village of Dhoreni at the only building in the vicinity to survive the quake—a small concrete-walled Baptist church. The rest of the village was pieced together with tarps and lean-to structures from remnants of previous homes. We requested to sleep in the church, and we let the head pastor, Rakesh Magar, know we were heading to Saurpani to deliver supplies and scholarships tomorrow to kids in need. He welcomed us with open arms and accommodations. Prem also mentioned that I was doing murals along the way, and Rakesh's eyes lit up. Upon his request, I painted a bird on the side of the church with a headlamp and the light of the truck shining the way. This time it was a dove to represent peace. Rakesh was pleased.

In America, it was Thanksgiving, and although late in the day, we mentioned to Prem we should have a little celebration. In the center of the village was a small market that served meals and drinks, so our band tumbled in. We ordered a round of dal bhat—a bowl of lentils and rice that is the national dish of Nepal. I eventually realized this was what we would eat for every meal for the next week and that was fine—it was delicious. After dinner, Isabella produced a small portable speaker and announced it was time for a Thanksgiving dance party. In the wet mud of the road, in a dark valley in the middle of the Gorkha district, surrounded by struggle, we played Michael Jackson hits. We sprang into jiggling and bouncing with the kids at the restaurant, using our headlamps as strobe lights, and the night sky was filled with laughter. Heads start peeking out of tarps, and one by one, villagers came out in their pajamas, carrying various instruments.

We turned off the Michael Jackson, and the villagers began to play traditional Nepali music with flutes, hand drums, and tambourines. Much to our delight, Prem's body came to life possessed with wrist-flicks, overhead rhythmic clapping, and spring-squats as he treated us to his signature Nepali moves. We danced into the night. We all laughed and had the most memorable Thanksgiving of our lives. Eventually, the band headed to bed, and we did too. It was quiet again, and reality set in for me—they had lost everything, but they had each other. Their lives would never be the same, but they could play music, dance, and experience joy in a spontaneous moment with strangers. I supposed that's the whole point of Thanksgiving.

I slept outside the church under a small banyan tree. I awoke to the smell of a campfire. As my eyes adjusted, I saw three men squatting around a fire, having a discussion. With them was Pastor Rakesh. I joined them, and through broken English, Pastor Rakesh told me that the other two men were also leaders in the community—one for the Muslim community and one for the Buddhists. They were all sharing use of the church building together, worshipping in their various ways. I found it heartwarming how they were sharing the resource as a community, even if it sounded like the start of a bad joke ("A Christian, a Muslim, and a Buddhist walk into a bar in Kathmandu . . . ").

We arrived in Saurpani early the next morning, and the village was alive with activity. A line of colorfully dressed women were shoulder to shoulder, passing large stones from a fallen building from one end of the village to another, like a human conveyor belt. Children with wheelbarrows moved supplies here and there, and the men were doing reconstruction. In the rubble of a fallen home, I saw familiar items poking out from the debris: a stuffed animal, a kitchen sink, and a small framed picture of a family. I sat with children in the school, and we drew for a bit, collaborating in my sketchbook or theirs. We eventually worked on a six-foot-wide mural of a butterfly together on the side of the school.

Prem gathered the town residents for a scheduled delivery. Isabella quietly explained to me what was happening.

First, Prem gave all of the kids coloring books designed to help them process what they had gone through and teach them what to do in case of future catastrophes. In the books, two cartoon mouse characters discuss what it means to lose a home and belongings, and assure each other everything is going to be okay.

Next, Prem called specific children forward. Isabella whispered that these were kids who lost one or both parents in the earthquake. Prem explained they would be given scholarships to attend school for the next year. Sweet little kids stepped up with tears in their eyes, sometimes followed by a grandparent or sibling. I cracked, and the tears dribbled down my cheeks.

Prem later told me about his first arrival in Saurpani after the quake. "I met a fourteen-year-old boy named Himal Khawash and his sister who had lost their mother in the earthquake. I remember his tears rolling down from his eyes, as it was still such a fresh wound. We sponsored the siblings for a long-term scholarship. Now Himal is a high school graduate and serves his community in the village." Another student, Milan Tamang, had lost his mother in front of his eyes during the disaster. When we arrived, Milan gave a welcome speech in gratitude of KarmaFlights' impact. The school had previously been teaching computer theory without actual computers.

Prem and his team took note and built a solar-powered information and communication technology lab for them to have the equipment they needed to progress. Milan remained on scholarship from KarmaFlights throughout his schooling and until graduation.

After five days of deliveries full of smiles and hugs, we arrived at Laprak, at 2,200 meters and the absolute epicenter of the 2015 quake. We were at the highest point in our journey, and it showed. It was significantly colder and exposed. The wind ripped through a village in shambles, with tarps flapping like a laundry line. Winter would be here soon, and the entire village of makeshift huts and tarps would have to deal with snow. We were surrounded by majestic peaks in a gorgeous panorama of Himalayan beauty. High above was Ganesh II at over seven thousand meters tall. I painted my last mural on the school based on the women I had seen rebuilding the country, and we delivered the last of our supplies.

Nick was finishing a plate of dal bhat and writing some notes in his journal when he had an idea. Something I had learned on this quest was that Nick was quite adept at seeing an idea through to completion. I had also learned that he was not here for himself but to be of service, like Isabella, Prem, and Jamie. He slid his journal across the table for us to review. Nick had figured out that if we requested a helicopter evacuation from our location, we could spend the rest of our trip budget on filling the incoming helicopter with supplies. We were only a few miles from Kathmandu by the way the crow flies. We unanimously supported this idea, of course, and instructions were relayed to the pilot to go into town and purchase as many coats, hats, and blankets as he could fit into the chopper, and we would pay for everything, including the flight.

It was a remarkably simple plan, and Nick's compassion and quick thinking made it happen. Despite the billions of dollars of donations that came from around the world, the village had received little support in the last twelve months because of its remote location. I heard more than once in Kathmandu that officials were hoarding rescue supplies in their own homes. With Nick's phone call, the incoming winter would be easier to bear for everyone.

The chopper arrived, and no space was wasted—the pilot was cocooned into his seat by bags of clothing busting at the seams. Everyone in the village received something, and as we unloaded, Q pointed at my own personal layers of clothing and said, "You can get more, you know." He was right. I took off my down coat and shell and left them behind, along with a knife, a hat, and gloves.

After a minor culture shock when we arrived back in Kathmandu, it was agreed we needed more dancing. Nick and I took a tuk-tuk down to Thamel Square and joined a throng of hundreds of twenty-somethings at the Purple Haze Rock Bar. Strobe lights flickered all around us, and it was definitely past our Gorkha district bedtime. On the stage was an excellent Aerosmith cover band. Nick and I hopped on the dance floor and were sure to include a few of Prem's signature moves.

Months later back in the US, Nick released a couple of small films online with partner GoPro, who promised to match all donations for the kids in the Gorkha. The films generated tens of thousands of dollars for KarmaFlights' ongoing efforts.

When I reached out to Prem and the Messengers years later, we were on the other side of a worldwide pandemic, and Nepal had experienced yet another devastating earthquake. Cosmic Brontosaurus again responded with support to their community with the confidence of a team that had been through this battle before. Beyond Nepal's borders, inexplicable wars raged on in various global hot spots. Most of us hide under the blankets of normal life, hoping it will all go away by morning.

But not Prem.

I didn't draw in my journal as much in Nepal as I did in Brazil. My energy was spent more on creating art that was left behind on cinder blocks and corrugated steel rather than returning with me on paper. This still felt true to what has now become my path: observe, find a connection, and draw. There was no rule stating the works had to come home with me, and they certainly had far more purpose there on the school walls in the Gorkha district.

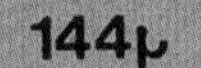

chapter

3

WHERE LIFE BEGINS

"We are the caribou people. Caribou are not just what we eat; they are who we are. They are in our stories and songs and the whole way we see the world. Caribou are our life. Without caribou we wouldn't exist."

– SARAH JAMES, GWICH'IN
(as told to RICK BASS in *Caribou Rising*)

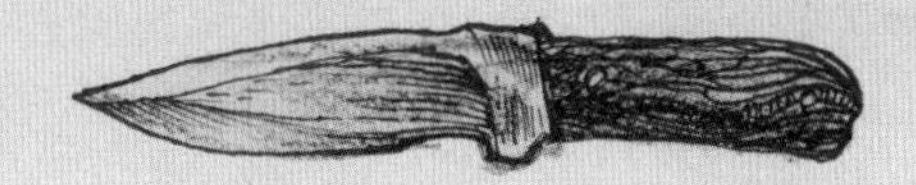

I gripped my knife in my hand as I fell asleep, still sheathed in its leather scabbard so I wouldn't stab myself awake. The six-inch blade was made of Damascus steel, and the handle was a walnut burl crafted by an Italian blacksmith named Patrik Francescato. It was a prized possession in my knife collection not just because of its quality and craftsmanship but because it was a gift from its maker. The little dagger spent most of its time sitting on a bookshelf as a decorative item I liked to admire on occasion, or perhaps slice an orange open with in a pinch. But for the past ten days, it hadn't left my side, paired with a bear-spray canister attached to my hip.

I was in my tent with a blindfold on to block out the twenty-four-hour sun of the Arctic Circle. Peter had the shotgun, or maybe it was Dan, but they were one hundred feet away, reminiscing by the fire. I wanted to be closer to the water to let the sound of the lapping waves lure me to sleep and stop thinking about bears.

The Gwich'in people often share this story: Twenty thousand years ago, this land was inhabited only by animals. Then came a great merging. Inside the heart of the Porcupine caribou was the spirit of the Gwich'in. Eventually the caribou released the humans to walk among the herd with an agreement that they would protect and sustain each other. For centuries, the Athabascans upheld this agreement by hunting the caribou for sustenance and survival but nothing more. The caribou herds have thrived and roamed the pristine boreal forest ever since, in the austere lands between the thirteen villages of the Gwich'in.

In 1980, President Jimmy Carter signed the Alaska National Interest Lands Conservation Act, which included protecting 19.3 million acres of pristine boreal and tundra wilderness, renaming it the Arctic National Wildlife Refuge. The refuge was not approved for oil drilling, but it has been a heated national debate ever since. Dan told me that the now-common abbreviation, ANWR, was initiated and promoted by the oil industry to avoid using the word wildlife.

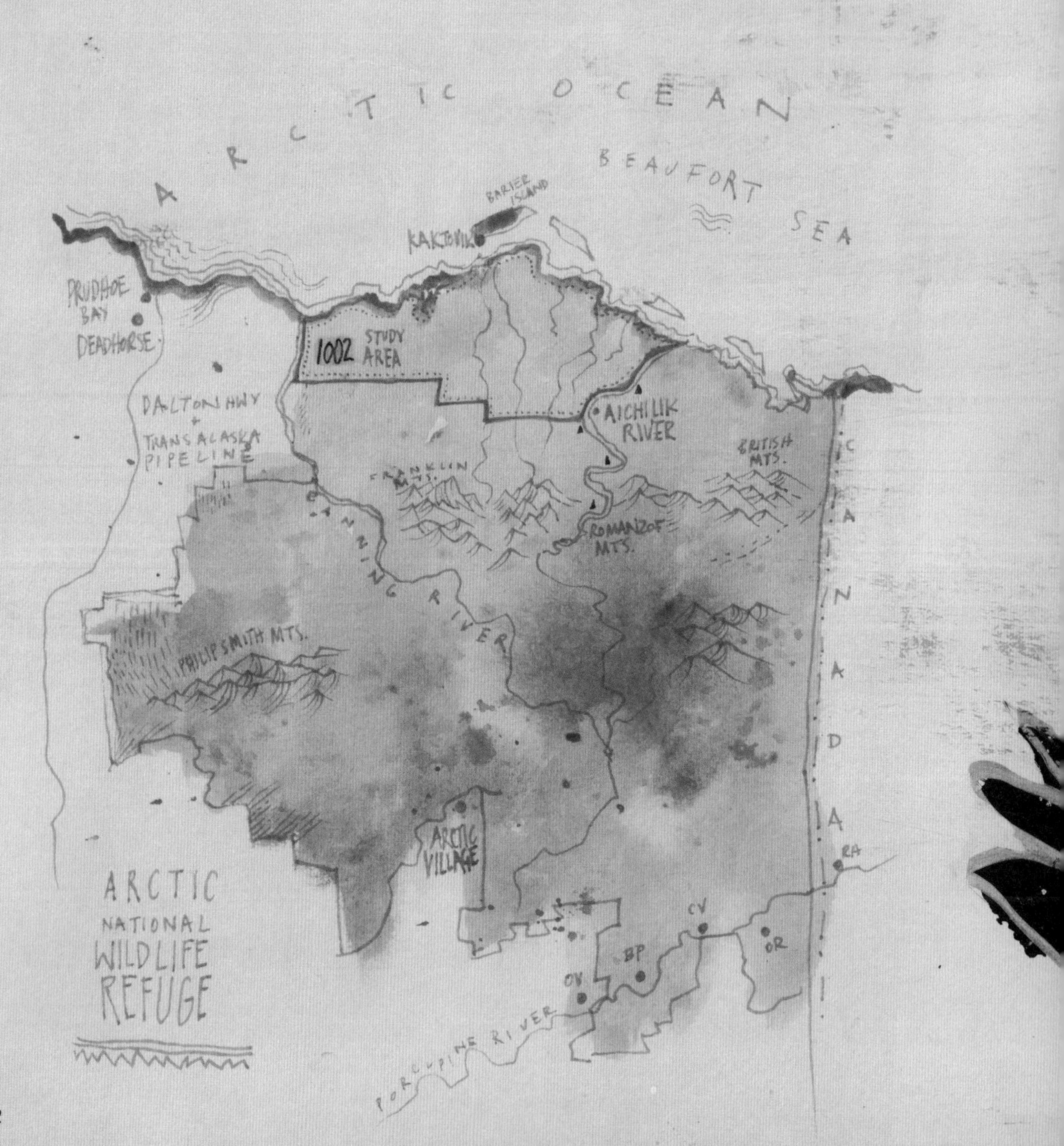

The Gwich'in have held firm that this sacred land is not to be invaded by development of any kind, let alone the introduction of drilling via a massive industrial complex on the banks of the Arctic, the home of their spirit ancestors and hundreds of species of wildlife. The North American Arctic is home to more than twenty-one thousand known species of highly cold-adapted mammals, birds, fish, invertebrates, plants and fungi, and microbe species.

On a cool June afternoon, I visited the office of Bernadette Demientieff in Fairbanks with Alaskan guides Dan Ritzman and Peter Elstner, as well as previous Sierra Club board of directors president Allison Chin. I hadn't been on a significant river journey since the Amazon trip two years prior. Bernadette was tall and olive-skinned, with high cheekbones and long, jet-black hair. I fully believed she was pulling my leg when she told me she was the grandmother of seven. She carried herself with ease despite bearing the weight of being a voice for the Gwich'in. As executive director of the Gwich'in Steering Committee, Bernadette interfaced with Congress, appearing in Washington DC regularly in public hearings, and was responsible for much of what the general population knew about Gwich'in spirit and concerns.

We were visiting Bernadette to understand the scope of where we were about to go—paddling the Aichilik River from the Brooks Range to the Beaufort Sea. Bernadette's parents, grandparents, and great-grandparents grew up in Alaska, infusing her with a deep respect for Indigenous values. She told her stories with conviction, giving voice to the legacies of thousands of years of ancestors.

"The Gwich'in and the Porcupine Caribou Herd have had a spiritual and cultural connection since time immemorial. We migrated alongside them for over forty thousand years. Our communities and the migration route are nearly identical. Our ancestors settled us so we can continue to live and thrive off the land and animals."

BERNADETTE DEMIENTIEFF

The battle for the Arctic Refuge has always been a complex balance of Indigenous, ecological, and human rights, as well as the rights of wildlife—similar to what I saw in the Amazon. Sure, the story that led us there was a social justice and human rights–based issue, but as I have found with most issues, social justice and human rights almost always reflect environmental concerns as well.

One of the biggest threats to the refuge has been and continues to be oil drilling on the coastal plain and its impact on the environment, wildlife, and Gwich'in people. The US government has often looked at Northern Alaska through a telescope rather than a microscope, observing the imagined vast potential but missing out on the details. Seeing through only one lens allows them to perceive only the short-term gain. After decades of political debate over access, Congress and the Trump White House collectively approved a drilling plan for the first time, and opened leases for drilling in the refuge. Bernadette proclaimed at the time, "Any company thinking about participating in this corrupt process should know that they will have to answer to the Gwich'in people and the millions of Americans who stand with us."

N
TO ARCTIC OCEAN →
BEAUFORT SEA
VABM 2537
HULA
1390
CAMP 1
2120
2725
5290
4390
MARIE
7910
Deliverance
McCall glacier
ISTO
9050
3920
LEFFINGWELL FORK
EGAKSRAK
ROMANZOF MOUNTAINS
JAGO
AICHILIK
BROOKS RANGE

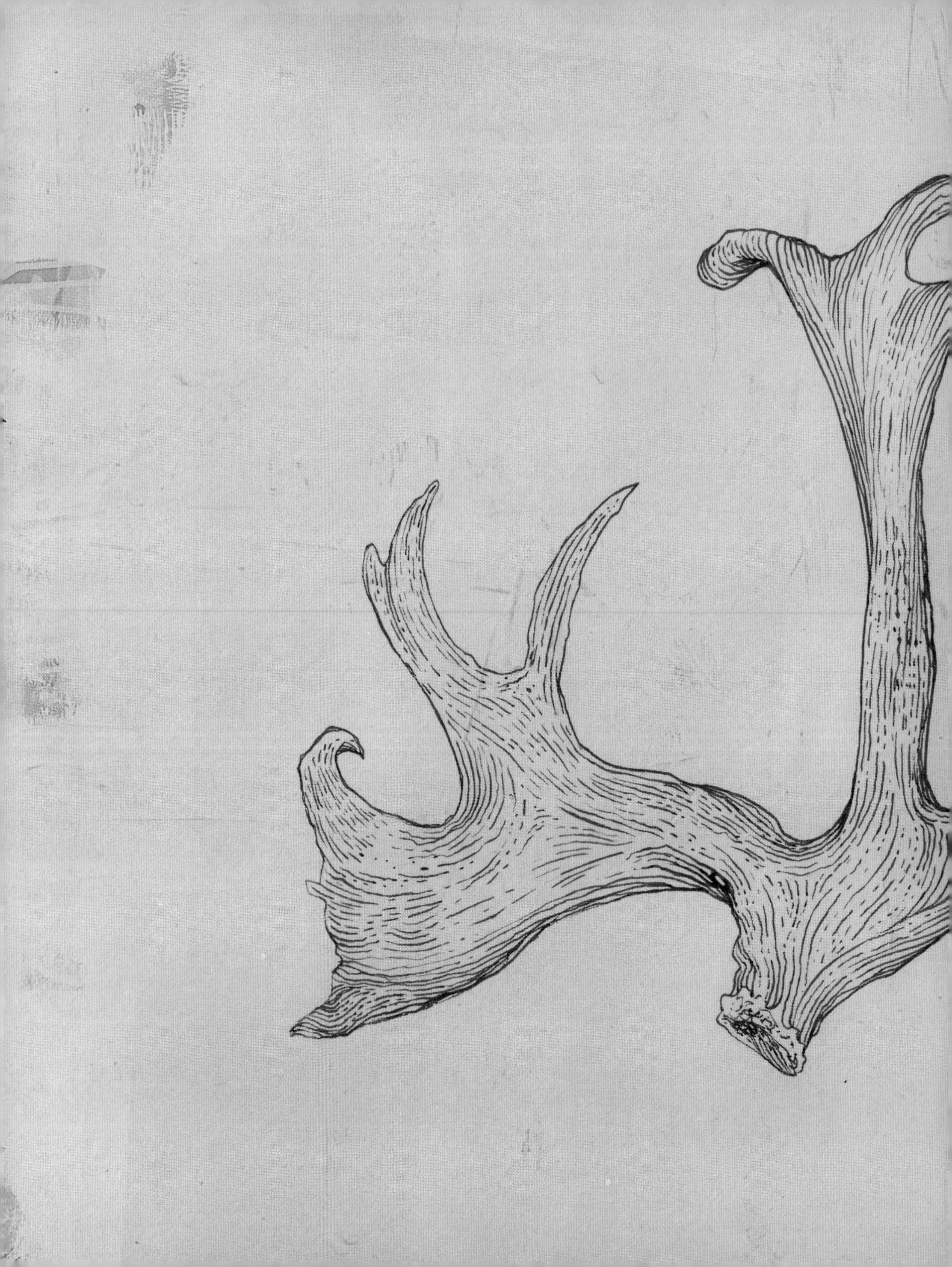

The Sierra Club hosts a select group in the refuge annually to have the authentic "wild Alaska experience," see it with their own eyes, and make their own conclusions. In 2017, they invited me. I arrived with an open mind, only mildly informed of the issues. My role was to paddle the Aichilik and produce a series of drawings that could be utilized to celebrate the wildlife and educate about the threats they face. By that point, I had grown into the art-journalism identity, and this was a dream gig. Dan and Peter's task was to make sure we survived the experience and perhaps secretly guide us toward falling in love with the landscape and the cause.

Within an hour of arriving at the headwaters, I was standing alone on a barren, windswept summit above the Aichilik River, looking east at the vast Brooks Range. To my left, jagged peaks tapered to nothing and eventually faded into a blue northern horizon. It was the kind of blue that I struggled to put a name to. It wasn't royal, cerulean, or indigo. Author Rebecca Solnit has a term called the "blue of distance," and I liked the sound of that. It was a perpetual nothingness at the northern end of our planet. The horizon was foggy, but out there beyond my sight was the North Pole. I had followed a prominent ridgeline on foot from our base camp, connecting a series of summits rising farther and higher. I had my knife and bear spray and trusty words of warning from Dan: "Do. Not. Get. Eaten."

If I couldn't climb to some glorious rocky peaks on this journey, I at least wanted to get my heart rate up on occasion. This was our last topography allowing for elevation gain, so I seized the opportunity. There was little to no vegetation growing on the ridgeline other than ground cover. Small Lapland rosebay flowers rocked their pink bobblehead blossoms erratically in the wind, letting you know "we are here!" But behind us and well beyond our view was the edge of the dense boreal forest extending over 105 million acres south of the coastline.

Along the two-foot-wide ridge, I came across a caribou antler shed. It was ragged, white, and roughly the size and weight of a tennis racket. I stopped and quickly drew it. Further on, I found another and then another, and at twenty-five I stopped counting. Some were larger, some smaller—I was clearly on a calving pathway.

With their migration based on weather or available food, caribou can travel through open tundra up to fifty miles a day. Much like migratory birds, they have a built-in compass and can even travel through unfamiliar terrain to reach their remembered calving grounds. Upon reaching their destination, they make the longest migration of any land mammal herd. Some scientists think that caribou might have a higher concentration of magnetite in their brains than humans do, a mineral that is known to assist with intuitive direction.

I pulled out my pen for a bit on a third high and windy perch, then began to make my way back to camp before dinner. The three miles of out-and-back ridge walking was my farewell hike to the dramatic Brooks Range. In the morning, we would start our journey northward.

We traveled in two four-person rafts captained by Dan and Peter. Dan was a lifelong Alaskan guide and the Sierra Club staffer who had first invited me to join them on the river to "do what you do." That shouldn't be hard in a dynamic landscape alive with birds and the constant sounds of movement coming from the Aichilik River and the spirit of those who call this place home. Peter was an expert naturalist and knew the waterways like the back of his hand.

They were both good company and had committed the bulk of their careers to a boots-on-the-ground understanding of this pristine wilderness before returning to modern civilization, where they were intent on protecting it. The voices of those of us who joined them on the water would assist in telling the story of the refuge in new and hopefully unique ways.

On the raft, I was sitting next to Allison Chin. A second-generation immigrant, Allison had broken all sorts of barriers in her role as president of the club's board of directors. Allison left a lucrative career in biotechnology working on cancer and age-related diseases. Whereas leading the Sierra Club may seem a far stretch from the lab, Allison saw a common theme to her work:

supporting the needs of others.

Allison's husband, Bruce, was along with us as well, but he had recently suffered a stroke and had trouble speaking. Allison bridged the gap, often answering the questions I asked him. I found all of them—Dan, Peter, Allison, and Bruce—charming, bright, and good company for the miles of river ahead.

Frozen alpine runoff lapped and splashed against our boat as we dodged the sharped-edged ice bank and exposed rock deposits. My eyes were glued to the horizon, shoreline, and sky at the same time, even as I tried to stay present in my paddling duties. I kept my binoculars around my neck for quick bird identification. I had yet to see more than a few small family groups or lone stragglers of caribou, but the birds were a constant source of awe and entertainment. Above us, Arctic terns hunted for char or grayling while preparing for their epic twelve-thousand-mile migration south next season.

The shoreline was damp, and we saw both grizzly and wolf tracks in the soft mud. Signs of wildlife were bountiful, but we saw no trace of human anything—no trash, no footprints, no fire rings, and certainly no massive industrial complexes with fences, dump trucks, and pipeline equipment. Because the sun was always up, we stopped for the "night" based on time. We unloaded the rafts, set up camp, and prepped dinner together. It was a treat to have a raft, and not just a backpack, and to carry more luxury items such as meat and wine.

There were three others in our group, spirited women in their sixties: Alice, Patricia, and Wanda. All were longtime Sierra Club supporters and birders like me, so we shared our finds. Dan and Peter answered all our questions and made sure we didn't miss a thing. Both were exceptionally patient with us as we paused to watch willow ptarmigan darting between grass hummocks. Glaucous gulls were prominent and a joy to watch too, but they could be a terror to others. Like mini pterodactyls, they often attacked other birds in flight and ate them in an explosion of scarlet-soaked white feathers.

Most nights after dinner, I excused myself from card games and walked away from camp with my fold-up chair and sketchbook to reflect on the day. When I sit to draw in these spaces, there's a process. I try to detach from whatever else I was doing previously in a couple ways. First, I'll read a few pages of whatever book I have along with me. On the Aichilik, it was Antoine de Saint-Exupéry's Wind, Sand, and Stars for the umpteenth time. Then, hopefully, I zone out in a kind of pseudo-meditative state just to quiet my mind with my sketchbook and drawing materials in my lap. In layman's terms, I'm chilling.

I harbor no pressure on myself to draw exactly what I am seeing like a traditional plein air painter. I carried photo references of local birds and wildlife I might see for only a fraction of a second. Tourist spots or local bookshops often had great waterproof foldouts for this. I was certainly not trying to perfectly replicate a grizzly that I saw as a blurry, lumbering dot through my binoculars. I often would lay down a loose watercolor shape, let it dry in the sun and wind, and then decide what I was drawing based on the negative shape remaining on the dried outer edges of the paint.

There is also a learned ability to ignore that comes with drawing on location. Ignoring the wind (forty miles per hour is not that bad). Ignoring an incoming storm (it'll probably pass). Ignoring the mosquitoes, gnats, and ants that tend to land or crawl on my hands (it could be worse). Oftentimes, there will be small brown smudge marks in my drawings. These are usually clues to silent mosquito murders. I've had watercolor paintings obliterated by rain and pages torn by gusts of wind, and pretty much every book has traces of spilled coffee, wine, or tea. Although they are, by definition, "sketchbooks," I rarely do much sketching in these travel time capsules. I draw. I write. I kill bugs.

I found myself obsessed with the idea of coming across a large caribou herd, but Dan tempered my expectations: "It's a fifty-fifty chance. I promise you, there's two hundred thousand out there, but don't get your hopes up." Dan went on to tell us about other riverways east and west of us, the enigmatic musk ox, over two hundred species of birds, and of course the covert polar bears. Listed as threatened under the Endangered Species Act, the polar bear population in the refuge is one of the most concerning in the world. With an estimated nine hundred bears, their population is almost half of what it was in the 1980s. Bears are continually leaving the ice pack due to increased melting and are moving inland. Their habitat is shrinking because of climate change, which, in turn, doubles the threat against them if their breeding and hunting space on the coast is invaded by large trucks and industrial facilities.

"My people are faced with losing thousands of years of culture for six months' worth of oil."

—FAITH GEMMILL,
GWICH'IN, ARCTIC VILLAGE
ALASKA

"Two strokes!" Dan instructed from the stern. Every paddle stroke sent the water into mesmerizing tie-dye swirls. I enjoyed the rhythm, and when the wind was blowing, we could go hours without chatting other than guided input. The constant gusting could be unnerving but certainly helped keep the mosquitoes away.

Three days into our forty-five-mile journey, the shore was now all ice. We were paddling down frozen hallways of aufeis with a full spectrum of blues peeking out through frigid riverside sculptures. The crystal-clear waterway pinched inward on occasion to ten feet in width, creating rapids that caught our attention but never felt threatening. We were in no hurry, and if we collectively wanted to stop for a trek out onto the ice, we did so. It wasn't like sunset was going to catch us. To walk across the frozen shore was like moving through another planet. It wasn't normal glacier ice but literally last year's waves stopped in motion, crunching and moaning under our feet. Rivulets of water raced beneath us and created a kaleidoscope of colors bouncing off sunshine and our own shadows. I saw azure, indigo, cerulean, cobalt, cyan, navy, sapphire, and ultramarine.

I would have liked to stay as long as it took to see everything the Arctic had to offer, but with a bush plane scheduled to pick us up in two days, we crammed as much as possible into every hour. I found myself wide awake at midnight, reading or drawing, sleeping very little. When will I ever be back here? I wondered. A faint fog rolled through camp, and a large family of caribou quietly passed by my open tent vestibule. I started to reach for my camera but changed my mind, and instead just observed and tried not to startle them. The buck looked directly at me, and I felt him try to make sense of the scattered colorful dome tents on the shoreline.

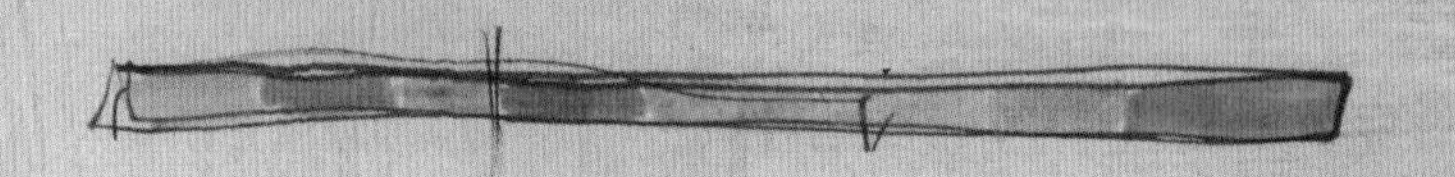

The herd faded south into the fog toward a zone called "Area 1002," based on Section 1002 of the Conservation Act intended for its protection. In February of 1999, Dan and a group of six other activists snowmobiled on the ice from Prudhoe Bay in witness (and resistance) to an audacious project that was drilling into the ice offshore to build a gravel island. BP called their $686 million plan the Northstar Project and expected to produce sixty-five thousand barrels of crude oil a day. To get the hot oil back on shore across the ice pack, they were burying a pipeline six feet below the Arctic sea floor, an approach that had not been tried before. Dan and his Greenpeace crew set up camp on an ice shelf just at the edge of BP's "no trespassing" zone—a mile of sea ice between them and the development. It was a balmy -25 degrees Fahrenheit in the Arctic as they erected their insulated tents directly on the ice.

Greenpeace had put significant energy into fighting this development for years in more traditional ways—through the courts and the court of public opinion. Nothing had worked, so Dan and the team used their human presence as a last-ditch effort to keep this monster from entering the Arctic Ocean like an oily Godzilla. The intent was to bring public attention to the offense, but once they arrived, there was very little they could actually do other than observe and film because of the atrocious conditions. A nasty storm kept them tent-bound for a week, then getting supplies into place was its own epic endeavor.

After one outing for filming, they returned to their rigid dome tent and found their locks broken and a note inside stating, "If this hut is left abandoned it will be confiscated," mysteriously signed by "the Alaska State Troopers." In response, Dan trekked back out across the sea ice to talk to the Northstar security team and try to understand who had written the note and why they had broken into the hut.

DAN RITZMAN

Twelve uniformed troopers quickly arrived and arrested Dan and two others for trespassing. They zip-tied their hands together and flew them back to Prudhoe Bay in a 737 full of oil rig workers. A $10,000 bond was placed for their release, and they were banned from the Arctic for a year.

Greenpeace sent out three more activists to the ice pack, but ultimately the project continued. Northstar Island still pumps oil from the Beaufort Sea to this day. Despite not necessarily being "successful," Dan felt good about the intent of their mission, both for the ecological impact and the Indigenous community at large. Because of the attention and disruption, other proposed offshore projects were shelved. Dan said, "In the United States there is no other place where communities of people are THAT connected to the land. The Gwich'in rely on the water and land for their subsistence. All of that together make it the most special place I can imagine."

As he stared out at the blue of distance, I could see in Dan a love for a place he has dedicated his life to. I am familiar with places that feel as close as a family member. They can be flawed, complicated, and difficult, but the comfort of familiarity keeps drawing us back.

THEN IT WASN'T

It was a river, and then it wasn't. The stream grew wider until it was no longer a river but the glassy estuary depositing us into the Beaufort Sea. We pulled over to the shore and marveled at the vast beyond of ice—sparkly, broken, and melting under a warm spring sun. For the first twenty-five years Dan was guiding here, this was frozen solid. Ever since the mid-1990s, melt has generally occurred earlier. A lonely ringed seal pup bobbed its head in rhythm with gentle waves. It side-eyed us as it headed south, then disappeared silently into a maze of frozen blocks.

We disembarked and walked the shore southward to visit an abandoned Inupiat whaling village. Remnants from the past were half-submerged in mud, and we quietly moved through what felt like a graveyard. Echoes of Indigenous life lingered here in the coastal wind as the climax of bird activity reached a fever pitch all around us. Loons, eiders, and terns filled the sky while pipers and a variety of sea ducks populated the shoreline. It was a lot to take in, and my head was quite literally spinning with my binoculars plastered to my face.

ARCTIC GRAYLING

Peter encouraged us back to the boats and proceeded to WALK us out of the estuary, dragging the boat behind him toward the ice a half mile into the sea. Apparently it was only shin-deep there, which was a party trick he knew we'd enjoy. Peter navigated past shallower deposits and guided us out into the depths where we observed massive floating chunks of ice separated from the main body. This pack-ice border was the polar bears' premium spring hunting zone, where they shopped for seals and fish. Peter kept the team shotgun ever ready and in eyesight.

Back in the rafts, we paddled to our exit point: a sandbar called Icy Reef on the edge of the northern tip of the earth. On this wasteland a mile long and one hundred feet wide, we unpacked and rebuilt everything for the last time, deflating our rafts and eating the best of the last of the food, like a climbing expedition—you always eat your best meals first, so as you have less and less with you, whatever it is that you eat is the best thing you've got. We started a fire from driftwood and told stories like all who had passed through here over the eons. It was almost midnight. Allison and I were now friends. Dan felt like a long-lost brother, and I would miss the flow of the river and his quiet ways.

We shared a flask of whiskey and some chocolate, and the spirits were high as we thought of our beds at home, our loved ones, and hot pizza. Allison and Bruce cuddled next to the fire on a log and watched it dance. Back in my tent with the hunting knife in hand, I eventually and happily passed out. Suddenly, a dark shadow passed over my face, and I inhaled quickly, terrified. My "protection" was useless if a polar bear was this close already. Fortunately, the shadow was Peter letting me know the plane would arrive in an hour, but more importantly, he wanted to know if I saw what was outside my tent. I rolled out of my cocoon and put on my glasses. There in the soft sod of the shoreline he pointed out a ten-inch-wide polar bear footprint, one of an endless path of them trailing off to the end of the sandbar. I muttered a choked "Uhh," and he busted out laughing, admitting, "Those were there when we arrived last night."

"Holy shit!" I barked, and my body instantly sprouted an orchard of goose bumps while we laughed together. The incoming pilot radioed us to say there was a beached whale a few miles south, so maybe the bear was headed there.

A short plane hop brought us to a landing strip in the traditional Inupiat whaling village of Kaktovik, and there we would catch a larger plane back to Fairbanks. Kaktovik might be the most remote neighborhood in all US territory. The village was a juxtaposition of run-down vehicles sitting next to brand-new ones, modern construction, and dilapidated shacks. Children ran in the dirt roads playing ball, and there was a community bake sale at the church. The population is just over two hundred people, and polar bears are a common sight in early summer—oftentimes at "the bone pile." Just outside of town near the airport lies a pile of bowhead whale leftovers the bears treat like a buffet. If it wasn't so grotesque you could almost call it modern art—bones stacked haphazardly like massive curved pick-up sticks.

Our plane took off, and as we banked south toward Fairbanks, the immensity of Alaska spilled out beneath us in all directions—a dark, dense, and endless wild with no discernible rectangles. To me, rectangles are the telltale icon that represents the human presence. Clear-cutting, neighborhoods, buildings, roads, maps, books, paper, even the laptop I'm typing this on are all rectangles. This book you are holding is a rectangle.

With my face pressed against the plane window, I saw no four-sided shapes for hundreds of miles in the largest undisturbed ecosystem in North America. The Alaskan Arctic is thriving in its fragility. A pipeline has no place there.

Days after President Biden was sworn into office in 2021, bidding for oil development in the refuge was put on hold as his administration canceled the remaining leases that had been auctioned. Bernadette took this all in stride with a skeptical optimism—there are still leases and potential drilling in other protected lands across the North Slope.

There were many outspoken defenders of the refuge before Bernadette, and hopefully others will come after. In fact, some say the refuge would never have made it through the 1980s protected if not for the inspired efforts of a part-time drummer from California. In 1987, Lenny Kohm visited northern Alaska and found himself inspired by the immensity, and was immediately concerned for its protection. Upon returning home, he began preparing a slideshow he could use to educate Lower 48 Americans about this special place and why it should be protected.

Others followed his cause, and eventually Gwich'in spokespersons joined his tour that lasted for two decades until his passing in 2014. "The Last Great Wilderness" was presented in libraries, university lecture halls, church basements, and other venues as an unorthodox yet effective voice for the Porcupine caribou and all other threatened species of the refuge.

Golden Eagle

Lenny wasn't the only one who used creativity to tell of the fragility of the refuge. In the 2005 film and book Being Caribou, writer and biologist Karsten Heuer and filmmaker Leanne Allison followed a herd of 120,000 for five months through this landscape, tracking their calving grounds and migration patterns. Creativity is at the heart of protecting this place. So who comes next after Bernadette?

When Trump returned to office in January of 2025 he quickly resumed the push for leasing and drilling in the refuge. Bernadette, Dan, and the other Arctic defenders continue the work to protect this place.

Before leaving Fairbanks, I asked a taxi driver where I could hear some local live music. She took me to the outskirts of town and down a darkened dirt road to a roadhouse bar. I heard the deep thumping sounds of an upright bass and some coyote-like hoots and hollers emanating from within. I entered the dimly lit bar via a log-framed doorway and vestibule housing local advertisements with an emphasis on taxidermy and truck repair. Inside it was packed elbow to elbow, with walls covered in Alaskan outback ephemera. It felt very "bar at the end of the world," and it kind of was. There were curved-up leather cowboy hats with aloft bird feathers, race-car T-shirts, a weathered glowing jukebox, and a bottle in every hand. I followed the sound of music out the back door into an open yard with a stage and string lights. At least one hundred people were dancing to the rhythm of a bluegrass band complete with a washboard, banjo, harmonica, and bass.

The demographics were a stereotypical portrait of Alaska—there were Indigenous people, folks in sleeveless denim jackets, thick-shouldered oil workers, and barefoot hippies. Regardless, they were all pulsing and laughing to the music together. Music is where humanity meets. An elderly First Nations man was in front of me on the outskirts of the crowd. I observed that his silver ponytail was threaded through an animal vertebrae. I leaned over and said, "Opossum?" while motioning to his adornment.

He looked at me with glassy red eyes and a smile. "Raccoon."

I attached my drawing satchel to the rear of my belt loop and joined Mr. Raccoon Tail on the dance floor.

"What values will we allow to shape the world we live within?"

CARLETON BOWEKATY
ZUNI, THE SALT LAKE TRIBUNE

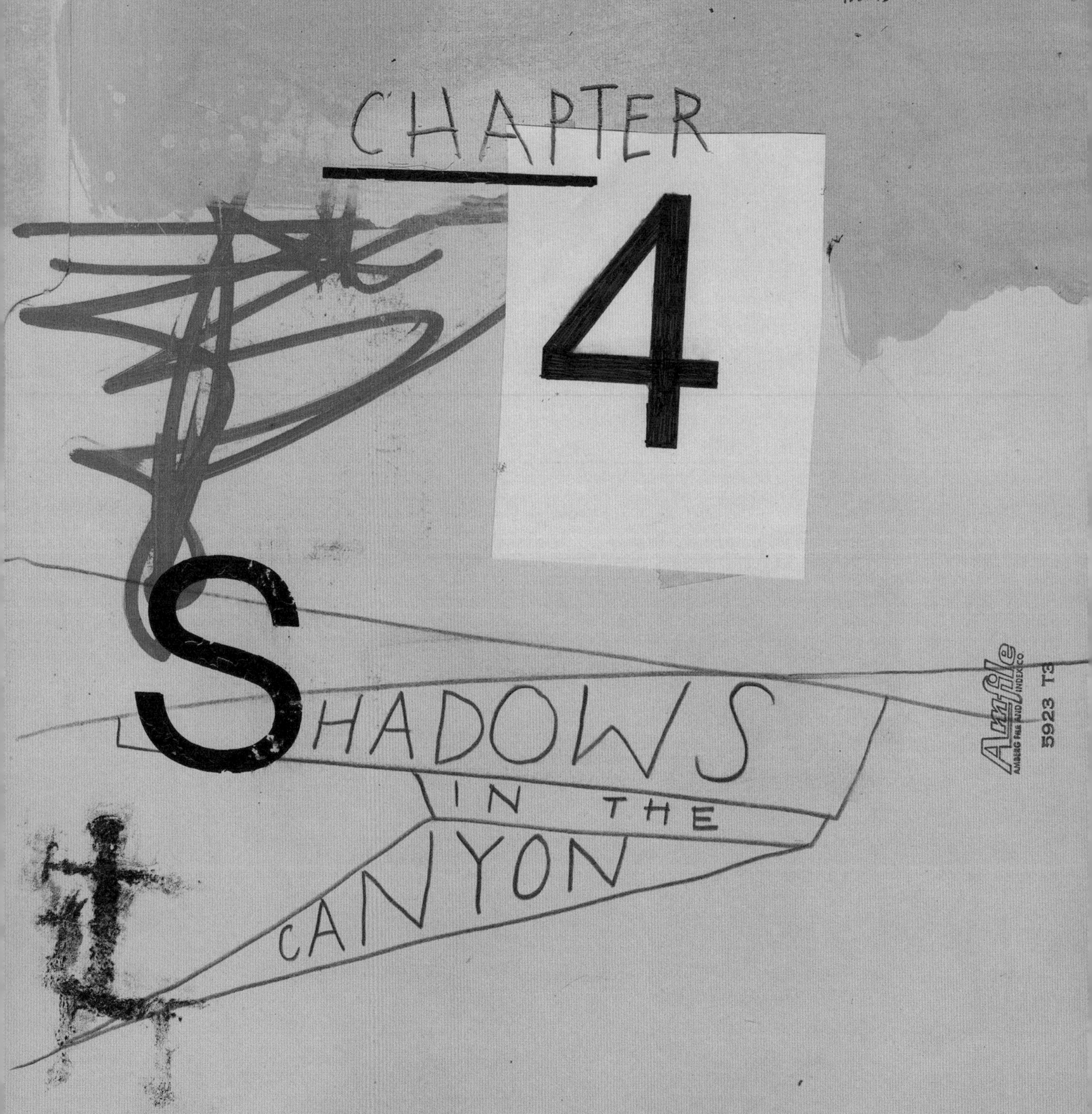

"Desk delivery is meant for lobbyists, not activists," a perturbed voice announced over their microphone on the Senate floor of Utah's capitol building. The room was stuffy yet ornately historic, robustly designed with large wooden pillars and decorative framing. A faint smell of cleaning supplies and wood sealant lingered in the air. I was seated in a radial bleacher in an elevated mezzanine, next to the Sierra Club lobbyist who had invited me, listening in to the day's proceedings. The delivery the senator referred to was from me: earlier that day, I had placed a hand-signed Valentine's Day card on each desk of the Senate floor. The front of the card featured my illustration of a roaring bear, and, in all caps, the words "UTAH, DON'T MAKE A MONUMENTAL MISTAKE. STAND WITH BEARS EARS." On the back I had added a handwritten "Happy Valentine's Day" on each one.

I watched, equally proud and repulsed, as the perturbed senator wadded up the card and threw it into his desk-side trash can. His intent, like many in the Utah administration, was to dig for oil in the sacred lands of Navajo, Hopi, Ute, and Zuni tribes—lands collectively known as Bears Ears.

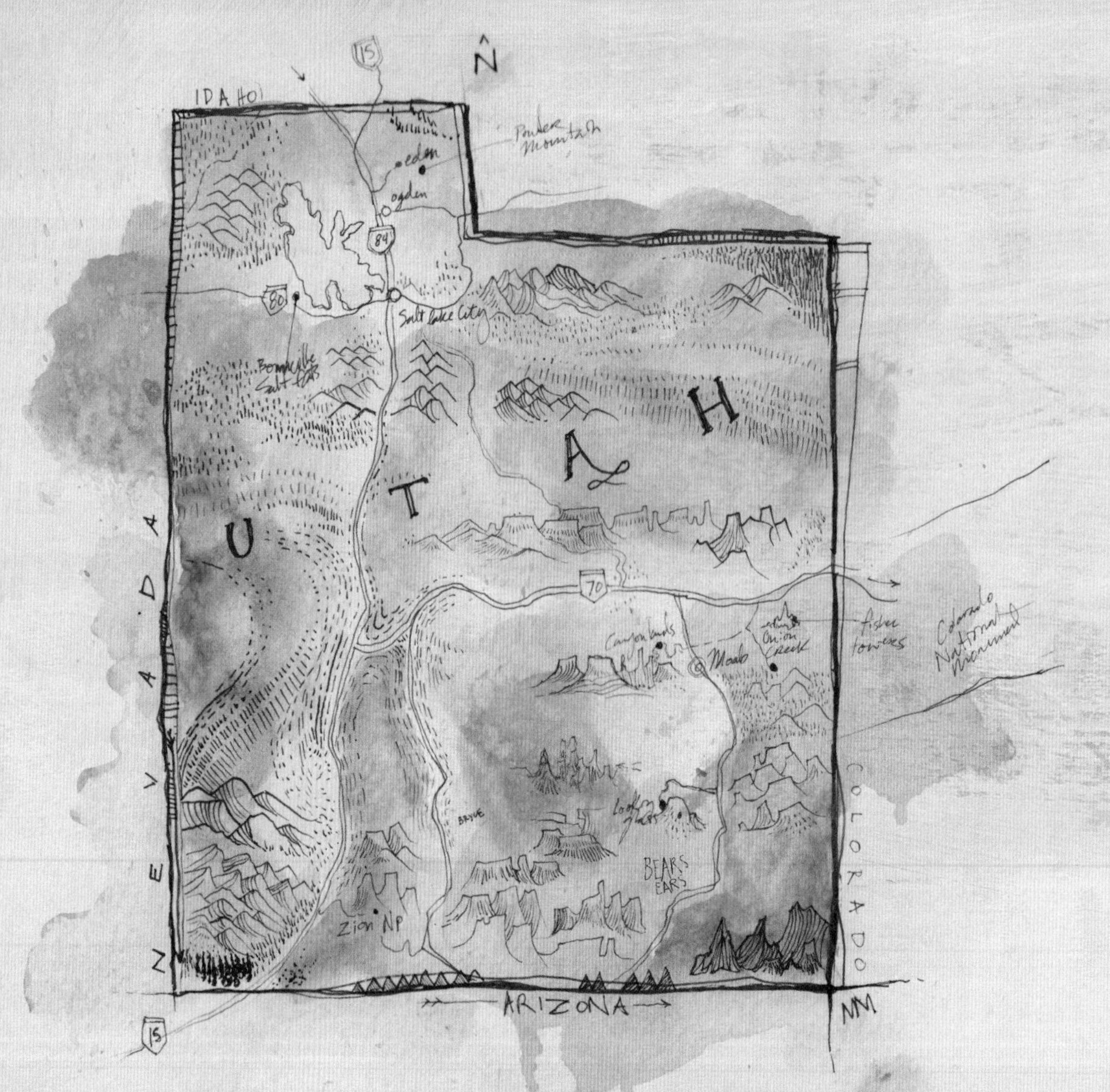

"Bears Ears" historically defines an area south of Moab and north of the Arizona border at the San Juan River. The vast, wind-sculpted landscape is a maze of canyons and corridors with the ghosts of Indigenous stories around every turn. It makes up a major portion of the Colorado Plateau where Arizona, Utah, Colorado, and New Mexico converge. Eons ago the region was an ocean basin filled with salt and then capped by ancient river sediments, sand dunes, and shorelines that would become the soaring naked cliffs we know now. The salt layer eventually bulged, creating cracks in the soft sandstone. Concave recesses in the walls crafted by wind and water gave a natural framework for dwelling spaces. The name Bears Ears comes from two flattened twin buttes representing a bear rising from the desert.

In 2016, President Donald Trump drastically cut the size of the Bears Ears National Monument, breaking it into two smaller pieces and leaving much of the land at risk of natural resource development.

Banding together in opposition from across the Four Corners states were representatives from the Navajo, Hopi, Ute, and Zuni tribes. Known as the Bears Ears Inter-Tribal Coalition, their mission is to "assure that the Bears Ears area will be managed forever with the greatest environmental sensitivity . . . where we can be among our ancestors . . . where we can connect with the land and be healed."

On my visit to the capitol, I wore a borrowed suit and tie and too-big-for-me shoes, and kept wondering how the hell I ended up here. Just twenty-four hours earlier, I was an hour north of Salt Lake City teaching a workshop in Eden, Utah, based on "Art in Activism." My experiences in the Amazon, Arctic, and Nepal were certainly part of the presentation, but I didn't feel like an expert, more like a fellow layman following a path of questions. At the end of the day, I read about the town hall meeting in Salt Lake City the next night, hosted by Utah representative Jason Chaffetz to discuss the future of Bears Ears with his constituents. I sprang into action and drew an image—the same image that would later land on those valentines—in response. I shared it online and offered it free for local participants to print out. Within hours, folks appeared on the local news protesting outside the meeting with full-sized posters of my bear drawing. All of a sudden, I was personally involved of the debate from afar.

Carleton Bowekaty, a former lieutenant governor for the Pueblo of Zuni, wrote in the Salt Lake Tribune in 2023, "Some individuals in the United States have acted as if truth no longer exists and that taking whatever you can grab will be tolerated. They acted as if merely stating something made it true, as if a person could violate the law and face no consequences. As Native people, we have faced this mindset before." Even though President Biden restored the scope of the monument in 2021, the future of the region remains somewhat tenuous due to legal challenges to the Antiquities Act and the potential for future administrations to again attempt changes.

Spruce and aspen dot the higher mesas peering far below into violet-shaded depths. Bears Ears is not only the sacred homeland of numerous tribes, but it also is the home of the mountain lion, golden eagle, and bighorn sheep. Prickly pear, yucca, cedar, and manzanita decorate the lower hillsides. Embedded stone and mortar granaries in the cliffs still house corn thousands of years later. This is not a forgotten wasteland ripe for development. This land is alive. The desert is a living shrine.

Government and industry are not the monument's only threats. For centuries, looters and vandals have ransacked the hills, taking what they can find. There is a story of a looter unearthing a deceased child wrapped in a cradleboard, buried beneath the soil. The looters dumped out the mummified carcass and took the cradle to be sold to the highest bidder.

After my capitol building stunt, I pulled off the suit and tie, changed into overalls, and jumped in a car with my friend Jonny Griffith as we headed to Ken Sanders Rare Books a couple of blocks south. Jonny was certain they'd be enthusiastic about a mural on the side of the building with the same imagery as the Valentine's Day cards. Ken was an old buddy of Edward Abbey's, and within minutes of showing Ken the image, I was pulling out a ladder and some paint. My friends Devaki Murch and Bridgette Meinhold came to help. In the frigid February air, our breath billowed clouds of condensation mixed with the mist of aerosol paint.

Four days later, the collective outdoor industry announced that they would pull their multimillion-dollar biannual trade show out of Salt Lake City, citing the low priority of environmental and Native concerns by Utah leadership. The people were not putting up with this.

Bears Ears National Monument
1.35 million acres

Proclamation #9558:
After years of petitioning by local tribes, in December of 2016, President Barack Obama declared 1.35 million acres to be protected as a national monument under the Antiquities Act. Although successful, this was only 75% of what was requested by the tribal council.

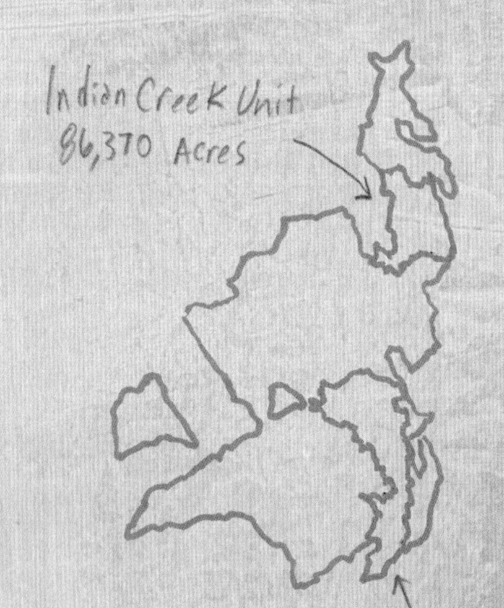

Proclamation #9681:
One year later in December of 2017, President Trump proposed reverting the protections into two small monuments to open the rest of the area for drilling and mining purposes.

GREEN RIVER
I-70
COLORADO RIVER
MOAB
191
CANYONLANDS
MONTICELLO
BEARS EARS

"The world is awake—except the canyon, its bottom still lost in fathomless black. Slowly the sun climbs higher. Deep down the purple shadows stir and swirl. Light breaks through the blackness, tinting the walls."

DAVID LAVENDER
RANCHER-TURNED AUTHOR

THE COWBOY

In the early 1980s, Matt Redd put on his dusty boots and cowboy hat to join his dad on a jaunt into Lavender Canyon, adjacent to Bears Ears and Canyonlands National Park. Robert Redd was looking for a spot to dig a stock pond for his cattle and needed an elevated bird's-eye view. This was nothing new for nine-year-old Matt, who spent his summer afternoons scrambling through a maze of cottonwoods and red stone tunnels to unearth arrowheads. Matt knew the best time to hunt was after a storm when erosion revealed all. He and his brother were living every kid's dream—they were the only youngsters on three hundred thousand acres of grazing land at the southern doorway to Canyonlands National Park and what would eventually become part of the Bears Ears National Monument.

As Robert and Matt scrambled through a jumbled conglomerate band of rocks toward the top of a mesa, ominous clouds broiled over the western horizon, accompanied by distant, rumbling thunder.

"Matt, grab some twigs and branches and toss them under the overhang o'er there," Robert instructed the young cowboy.

With a darkening sky, together they summited the mesa to survey the land below. Robert pointed out features in the land that would or wouldn't make for a good reservoir. To the west of them, sprouting like skyscraper tulips from the landscape, were the iconic three-hundred-foot tall North and South Sixshooter Peaks—prominent towers with convincing handgun barrel features. To Matt, they seemed as mysterious and sacred as the Vatican holy places. It wasn't long before the sky cut loose and the Redds ran for cover from a cavalcade of hail and lightning. They scrambled off the mesa and down to the overhang where the wood stash awaited them. Robert quickly lit a fire to keep them warm until the storm passed.

The rain continued as they dried out, laughing at their situation and stoking the fire. Then a roar that sounded like a train came rolling over their heads—the buildup of water on top of the mesa gushed violently over the lip of their hideout. In moments, it went from terrifying to glorious as the sun, waterfall, and fire created a kaleidoscope of shimmering color in their hideout. Young Matt was in heaven.

INDIAN CREEK
Sunday, August 9th
N
super bowl
HART
POINT
21
N →
S
BRIdger
JACK
MESA
Donnelly canyon
Titus
Canyon

Eventually, the rain subsided, and the Redds made their way out of the canyon after crossing a gauntlet of now-flooded creeks. As darkness fell, they walked atop a large fallen cottonwood over the last runoff and made their way out.

Matt and Robert arrived at home soaking wet with sheepish grins. Robert's wife, Heidi, took it in stride as just another day at the Dugout Ranch.

"Get dry. Supper's ready."

A fourth-generation rancher, Robert had purchased the 5,200-acre ranch in 1962, and the Redds cut ties with the outside world, having no phone or electricity. In 1997, after a divorce and many years of challenging finances, the Redds realized they would have to sell the land. Rather than parcel it out for condos or the most beautiful golf course in the galaxy, Heidi came up with the idea to sell it to the Nature Conservancy for a song, as long as she could stay on the ranch and manage cattle for the rest of her life. The deal was set, and ever since, the Nature Conservancy has utilized the Dugout and surrounding lands for the Canyonlands Research Center, which explores the interactive effects of climate change and land use to arm decision-makers with new information for adapting to challenges such as grazing, invasive species, and recreation impacts.

After finishing college with a double major in photography and creative writing, Matt returned to the ranch to keep the cowboy dream alive and work alongside the Nature Conservancy. Heidi has since become a western icon. She was inducted into the National Cowgirl Museum and Hall of Fame and is the poster child of surviving the gentrification of the West. "I've always figured cowboys and cowgirls have the best jobs," she noted in a late 1990s interview. "They probably have the lowest wages, but they don't have to wait till the weekend to enjoy life."

When asked if selling the ranch to the Nature Conservancy felt like a loss, Heidi said, "I never owned the Dugout Ranch; Heidi Redd never owned this place. This place has owned me from the minute I dropped into this valley, and it has been my absolute love to keep this place free of development in its beautiful natural state."

No one I've met in my life embodies the American spirit, or what's left of it, like Heidi. While the rest of us dingbats haggle over parking spaces and gas prices in rotting cityscapes of asphalt, Heidi is out there tending cattle, wrangling horses, and watching decades of sunrises and sunsets in a heaven we barely know.

My hands were wrapped in two-inch athletic tape like a mummy—up over the knuckles, under the palm, a loop around the wrist, and back up toward the fingers in a figure-eight pattern. Then I wrapped smaller strips around the lowest portion of my fingers between the first and second knuckle. My hands looked like I was preparing for getting into the boxing ring, and it wasn't far from the truth. I was climbing in Indian Creek, Utah, a popular section of Bears Ears and home of the world's highest concentration of "splitters"—that is, cracks that are laser-cut fissures ranging from half an inch wide to accommodating an entire body. My tape gloves were the only thing between my skin and the rough stone, minimizing the wear and tear on my hands because the only thing better than climbing is MORE of it.

To ascend a crack is like climbing a suspended rope in gym class, except there is no rope—you are climbing what ISN'T THERE rather than what is. For many climbers, this is a grotesque way to climb—jamming and crushing soft body parts into vacant spaces—but for those who embrace the pain, Indian Creek is easily the world's best playground for it.

In "the Creek," crimson walls of Wingate sandstone splinter up from hillsides decorated with tumbled boulders fallen from the cap of the Kayenta strata two hundred feet above. Shadows of violet and indigo reveal corners, cracks, and nooks along the seemingly endless miles of rusty cliff line.

As a climber, I squinted to pick out the most attractive cracks, and as an artist, it was practically impossible not to mimic the geology with pen and paint as a means to understand it all. In the valley below, family clusters of cottonwood and aspen bordered the creek snaking its way through rolling yellow hills. Cattle dotted the landscape on the historic Dugout Ranch.

The motion of crack climbing is a full-body experience that involves squeezing limbs and digits into rows of shark's teeth between two featureless skyscrapers. When I actually type it out, it sounds less than enticing, but when you find your rhythm, it is certainly intoxicating. Leaning back on perfect hand jams, I focused on keeping my heart rate down, which in turn kept my body temperature down. I focused on staying relaxed—hand-foot-hand-foot for one hundred feet until I arrived at the anchors, a little less than half my rope length.

"Take," I yelled down to Renan Ozturk, who was belaying me on another perfectly straight line. He pulled the rope tight in his belay device and lowered me back to the ground on our spindly, bright colored rope. I landed with a big dumb smile and sputtered, "It's just so good! It's easy to forget when you haven't been here for a while." Renan and I met years ago amid the hot winds of southern Nevada. He was a career dirtbag at the time, gaining fame as a solo climber and artist while living out of his friend's beat-down Toyota Corolla named Silver Lightning. He ate, slept, and breathed rock. His hair sprung from his head like a giant yucca, and grit had taken up permanent residence under his fingernails and behind his ears. Climbing was his daily bread, but his actual sustenance often came from the dumpsters behind the closest grocery store.

I bought a six-foot-wide painting of Zion Canyon he had rolled up and smashed into the trunk of the Corolla. Whereas my arsenal is often quite minimal, Renan paints with whatever he can get his hands on—alcohol ink, marker, acrylic, oil, gesso, blood, watercolor, you name it—all mashed together like a cake. The painting is a bird's-eye view of Zion Canyon, focused on the Watchman feature from the park's southern entrance. It has the spastic energy of a Van Gogh, with the desert colors amplified in saturation. Darting hot pinks and fleeting ochre yellows highlight Renan's signature Fauvist-inspired line making. In the years since then, he has risen to fame with multiple film credits to his name as a trusted producer of National Geographic productions. His most notable film is Meru. "Oh, that guy," you might be saying. Yes, him.

Renan no longer lives in a Corolla, but part of his charm is that he carries that dirtbag spirit with him to remind him where he came from. He still is dirtbag beneath the fingernails and crust of success. I handed him the rope, saying, "Your turn, 'Nando," and took off my climbing shoes.

RENAN OZTURK
JAMES "Q" MARTIN

James Q took over belaying Renan as I looked out over the Dugout Ranch. Indian Creek had become somewhat of a home away from home for Q, but not only for climbing. Years ago, he met Matt Redd before an ascent of the South Six-shooter tower that Matt had seen from afar as a child. Matt had only flirted with occasional climbing but was finally ascending the South Sixshooter he had admired his whole life, with a friend he shared with Q. Matt and Q connected that day and eventually made a pact: Matt would teach Q how to be a cowboy, and Q would teach Matt how to climb.

Q started his career with a Canon film camera and a one-way ticket to Patagonia in the early 1990s. With a similar dirtbag ethos as Renan, he started shooting landscapes and then mountain athletes while living from a vehicle or a tent. He eventually landed in the magazines and catalogs popular in the day. For decades since, he has followed the siren's hustle of freelance life, taking opportunities that come his way but still making time to pursue his own adventures. At this point, I have taken more international trips with him than my own family. Based out of Flagstaff, Arizona, he keeps his bags permanently packed in a small shed outside his five-hundred-square-foot studio home.

He has traveled the entire planet with his camera since those early days but eventually found just shooting stills wasn't enough. Everywhere he went he saw the need for stories to be told that make a difference, and his camera was already the perfect tool. Expanding his skill set, he learned the art of filmmaking to amplify causes of environmental and humanitarian concern.

As the sun set on the fractured canyon, Renan, Q, and I headed to the Dugout Ranch to rendezvous with Matt and writer Craig Childs for dinner. Craig is a desert authority on a variety of 'ologys—archaeology, paleontology, geology, the list goes on. Among his numerous titles, his book House of Rain put him on the map. In it, he tracks the "unsolved mystery" of the fate of the Anasazi in the Four Corners region during the eleventh century.

In person, Craig is a paradox. He exudes youthful and ancient energy simultaneously, like a Roman statue swimming in a fountain of youth. His white beard leaps from his face like a cholla cactus, with blue eyes that spin like the ancient radial petroglyphs he obsesses over. He writes about mammoths, ancient cultures, and finding water where it isn't supposed to be—but not from a professor's office full of books. Craig GOES and becomes his stories. He walks the path of the Pueblo ancestors and cliff dwellers. He sleeps in the caves he writes about.

I imagine if you sliced CRAIG open Like a cake, SAND and arrowheads WOULd pour out. HE Doesn't write about the desert— he reverberates it into being. Craig is a walking echo.

The four of us came together over a shared appreciation and reverence for a place called Bears Ears, but we didn't all share the same perspective of it. Matt saw legacy and labor, Craig saw history, and the climbers saw lines to climb. We each tried to share the canyon with one another through our own eyes.

After fifty years of life in the canyon, Matt had explored all but a few remaining nooks and crannies on horseback. Capitalizing on our climbing and descending skills, he was hoping to navigate to a spot he'd never been to—a remote, grand mesa bordered by twisted tunnels like a castle's moat. It was nothing particularly noteworthy, just somewhere he hadn't been, and that was enough for him. Our objective was purely following Matt's curiosity. Slowly. Looking at the map, I didn't think it seemed easy, but we all brought something to the table. Matt brought curiosity and navigation, Craig brought water locating and petroglyph sighting, and the rest of us knew how to climb up and get ourselves down rolling canyons.

We scattered our supplies about on the Nature Conservancy office floor in preparation.

"Who's carrying the trowel?"

"How long should our rope be?"

"How much cheese is too much?"

I enjoy the act of preparation for a walking journey with a team—paring everything down to the essentials. The decision-making is practically ceremonious. We lay everything out like the inside of a cat's kill den and divide the items until each person's pack is full. My guilty pleasures include an umbrella for shade and rain, a fold-up chair for drawing, and my binoculars. I always look forward to being mocked for my umbrella and then, without fail, being asked to borrow it. One unspoken rule for multiday trekking is to never pick up your travel partner's pack out of curiosity. It's bad for morale to know if it's heavier or lighter than yours. If you all packed as a team anyways, you KNOW whose is the heaviest. I'm pretty certain Renan's pack would have won the heaviest prize with his art supplies and camera. But I don't check. Ever.

It was February 2020, a year after my state capital appearance, and as we started to saunter into the cool shadows of juniper and manzanita, we discussed this thing called "Covid-19" we had been reading about.

I shared from my infinite pool of wisdom. "I'm sure it'll blow over in a few weeks."

Craig quipped, "I think that's what they said about tuberculosis."

Little did we know we were weeks away from shopping for face coverings and taking unpleasant nasal tests. We pinballed down switchbacks, laughing and backslapping like naked emperors, entirely unaware this would be our last group socialization for months or maybe a year. Without cellular connection, we willingly let go of what was happening in the rest of the world.

Matt led us into the canyon on the outskirts of Canyonlands National Park. Although we all were connected in various ways, this trip was our first time together as a group. It was a unique journey for each of us: Matt was not there to gather lost cattle. I was not there to create for any specific project. Q and Renan were not there to make a film, and Craig wasn't there to tell a singular story. We had come together just to be. After a few miles, we set up camp on some flat boulders at the edge of a dry arroyo. As we settled into our sleeping bags, the stars splintered across the sky above us in a maze of blue-and-amber-colored pinholes. To give the cosmos a chance to truly perform their greatest concerts, we must visit them in their proudest auditorium—the desert.

Maybe it was a "canyon" on the map, but I quickly recognized that we were walking through an Indigenous artifacts museum. I generally consider myself one who treads lightly, but I quickly started fantasizing about floating through these corridors like a feather to avoid any threat of impact. Puebloan petroglyphs and pictographs were peeking out of practically every soft-shouldered rise, raw and fresh as if they were created last week. Craig pointed out the blatantly obvious AND not-so-obvious signs of past life.

The twisted mesas were as worn and splintered as an elder's hands, whispering a history for those interested in listening. Out there everything is held together by a mystic brittle glue. If I were to define the Bears Ears region by a singular word it would be "sensitive."

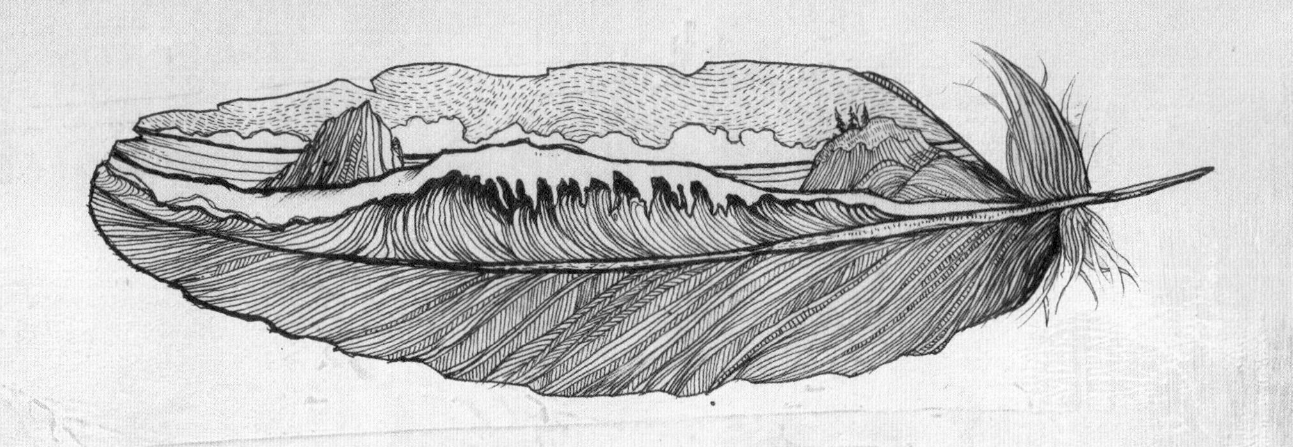

We chose to roughly follow a line of latitude rather than zigzag left and right in an effort to see things we wouldn't normally see. I found a chimney in a slot canyon that we would all be capable of climbing and dragged a nine-millimeter rope up the toughest section to bring up the team. One by one we humped our way out and over, squeezing and pressing through a slot approximately half an inch wider than Craig's barrel chest. We followed a rounded bridge of rock over to the next canyon to rappel and then repeat. Surrounded by hoodoos of rock hundreds of feet tall, I pointed out features like a child would: "That looks like an ice cream cone." Q positioned himself to be licking it in a photo. We were reverting to childlike versions of ourselves at summer camp. Then Craig pointed out another massive panel of rock art we had overlooked and ancient grain rooms high off the desert floor that we weren't capable of reaching even if we'd wanted to try.

The early inhabitants here were innovative engineers; they used tree trunks carved with footsteps to lean into walls of rock with shallow peg holes to keep them steady. One theory is they left the canyon after many seasons of drought, and that made sense to us as we struggled to find water to filter, but as Craig promised, it was always there. The mystical space we were trekking through could have backhoes, dump trucks, and pipelines one day if the decision-makers in Washington don't commit to Bears Ears' existence and protection.

On the fourth morning, I got up early, grabbed my drawing satchel, and headed out for a walk. Everyone was still asleep. It had been a cold night—a snow flurry came through around midnight, and the team quickly pulled out micro-tarps and bivy sacks. Luckily I had my trusty umbrella to put over my head. The snow highlighted small edges as I wound through spiraling corridors past obvious Indigenous dwelling spaces. Pottery shards poked up through the mud as voices from the past waved hello in the form of clay handprints accenting the walls. Above me somewhere a raven CAWED at the rising sun. I CAWED back in a throaty chortle. To my delight, we cawed back and forth for a couple of minutes as I got closer and closer to where the calls were coming from, snaking under twisted juniper limbs and scrambling up sculpted rocks.

"CAWW." "CAWW." "CAWW."

We were having a real conversation.

I mantled up onto the highest point in the area, and there was my raven—it was Craig sitting cross-legged and watching the sun come up. "Oh, was that you?" he asked. We both had a good laugh and quietly prided ourselves on the quality of our raven-speak being enough to trick each other.

"MAN, I thought I was having a real connection."

"Oh, you most certainly were!" Craig the raven said.

I sat next to him to draw as soon as my hands warmed up. As it often had, the topic led to our kids. We lamented and swooned and cried and laughed, experiencing the full spectrum of being dads. Parenting is a labyrinth. When our kids' world is on fire, we have to be water, not gasoline, with no control of which liquid they THINK we're delivering. Matt, Craig, and I each had teenagers a couple of years apart at the time, so we took turns prodding each other for advice and licking our wounds. We jokingly started calling our trek "the men's retreat."

Up and over, up and over, we made our way west on a sandstone roller coaster. We made a pact not to share the path we took publicly, but suffice it to say, it was not one we could ever repeat twice anyways. Every move was spontaneous, with a looser grip on our destination.

At one point, Craig's eyes darted from the left side of a small canyon to the right, as if listening to ancient whispers. High above us were now the ever-present pictographs. The canyon narrowed to a pinch to our left, then veined out into numerous side canyons to the right. Craig set down his pack and scrambled to a Volkswagen Beetle—sized boulder. The rocks were telling him something. He sniffed around like a coyote tracking a rabbit. He reached his hand into a hole under the boulder and confirmed to us "yep" with a knowing nod and bouncing eyebrows. For all we knew, he was pulling out a rattlesnake by the neck, but remarkably, it was the base remnant of a woven Pueblo basket. We all admired it, then Craig slid it carefully back into place. A gift from those who came before and to those who come after.

In 1896, rancher Ed Turner and his nephew Mel were out here wandering the desert much like we were, but on horseback and with less sophisticated bedding. They were searching for some lost stray cattle and scrambled up to the abandoned ruins of a home in the rock for a better view. Here they found a large, intact pot holding what has become known as "the Telluride Blanket." Alternating in stripes of brown, red, and white cotton, the blanket is one of only three known perfectly preserved prehistoric blankets. Experts say it was crafted during the time period of 1041 to 1272 CE. Luckily, the Turners protected it, and it has made its way through the Colorado Plateau in many different hands since, finally arriving at the Telluride Historical Museum. But how many artifacts in those days were shoved in forgotten corners of a garage or destroyed? As visitors to these sacred places, we're left with a few choices: take it and keep it, leave it and hide it, or pass it on to those who can protect it better than we can. Craig scratched at his beard and told us it was a "case-by-case basis."

On our final morning, we each rose early from our hidden holes in the sand, knowing we would regret it if we didn't soak up every remaining moment. Q boiled water and prepared coffee for the group as spiraling wisps of steam curled into the haze of his headlamp. We converged together on a bare slab of purple rock, still half in our sleeping bags, waiting for the rising sun to thaw us into motion. I certainly hadn't achieved Craig's level of awareness but felt I was learning how to SEE the desert, not JUST observe it. Anyone can observe. I observe a parking spot at the airport, the rising price of lumber, or a daffodil in spring, but to see takes effort. To see is to observe more than what our eyes can comprehend. Seeing is a multisensory experience, taking the full scope into consideration with the intent of understanding.

The pessimist says we see what we want to see. I believe we see what we seek to understand. In Bears Ears, I sought to understand what it was that we were all bent out of shape trying to protect from the bulldozers. Sketched out in ochre and ghosts, I saw the actual legacy of the land of the free.

After coffee, we packed up our camp, but without talking about it, we each left out our creative tools. Q and Matt went in different directions with their cameras and journals. Craig's writing hand spastically attempted to bottle whatever windstorm of words was swirling in his mind. Renan and I scrambled on top of a boulder with a rolled canvas he had carried with him the whole trek. We painted and drew together with arms darting in and out of each other's like a four-armed machine. Ochre, sienna, and turquoise bounced between us in a rhythmic dance of making.

On a crimson wall behind us were three sun-bleached pictograph masks that had survived wars, drought, and the modern age while looking out over a tamarisk-choked ravine. I hoped they were entertained by our little performance.

We created for the sake of creating. We stretched our tired limbs each morning and felt sand in our toes and sun on our skin as dust dried into beads of sweat on our faces. If we wanted to take a nap in the sun, we said so and did so. We were just shadows in the canyons like the passing of a flock of overhead ravens.

"For artists of all stripes, the unknown, the idea or the form or the tale that has not yet arrived, is what must be found. It is the job of artists to open doors and invite in prophesies, the unknown, the unfamiliar; it's where their work comes from, although its arrival signals the beginning of the long disciplined process of making it their own."

REBECCA SOLNIT

- A FIELD GUIDE TO GETTING LOST

CHAPTER 5

The Places That Scare you

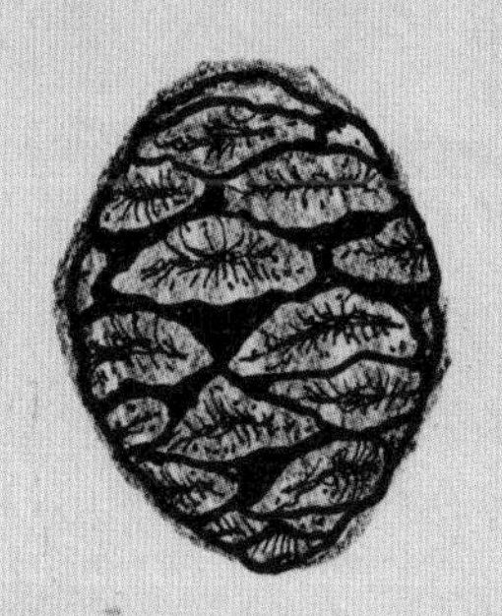

From my elevated perch in the sycamore, I could see the police searching for me. Their flashlights darted between houses, beneath bushes, and under cars, but oddly never upward. My heart was beating like a school marching band through my chest, but apparently they couldn't hear the drum line down below. Surely they'd see my toes over the edge of the limb, like a lineup of baby birds. The officers scoured the neighborhood and interviewed each neighbor. My poor mother cried. But still I didn't budge from my lush green camouflage.

Up in the tree I was a kid trapped in fear. How furious was my dad going to be that I had broken the glass top of his home office desk? The desk's platform was a hollow paneled door, reclaimed from who knows where and elevated by painted cinder blocks in the corner of my parents' bedroom. It was practical and frugal, like my dad. The offense was an innocent accident, but that didn't matter. In my mind, I had committed a felony and would be punished. Maybe prison.

I knew I couldn't stay up there forever, and eventually I felt less guilt about the desk and more about the anguish I was causing my parents. At least I could repair one of those infractions by simply showing up, apologetic and late to dinner.

Whatever anger my dad had about the desk was cooled by the relief of my return when I entered the house with my head hung low. My guilt and empty belly were punishment enough.

That was a lifetime ago now, but it still stands out to me as an early moment when I felt safe in a tree like Shel Silverstein's famous greedy boy. I remember being above it all for a moment, and I suppose trees still offer me that escape today—not from my dad, but from the madness and overwhelming speed of the world.

Growing up without internet, cable, or even a VCR, my brothers and I found our backyard to provide immediate access to entertainment and an escape. Stepping out the back door was like exiting the hatch of a spaceship. We used our imaginations without devices to inform us about the world.

We caught crawdads and lizards and would walk miles into the forest before they became suburbs again. It was western Missouri in the 1980s, and though our house was quaint, the yard in our minds was immense. It was a place to play, explore, and yes, hide. I can still sense the rough bark against my back, the sharp edges digging into my bare feet, and ants running over my toes.

My mom raised us to go outside in any instance:
Bored? "Outside!" Obnoxious? "Outside!" Too hot? You get the picture.

Just a couple of miles away at my Grandpa Bill's house, an oak sapling he called "the Jeremy Tree" grew in the front yard. Grandpa "Tiger" Bill planted it front and center of their home when I was born to celebrate their first grandchild. One of my strongest memories of him was his baritone singing voice, which I loved to hear. We would drive down to the lake in his pickup truck to go fishing, and Grandpa Bill would push in a Johnny Cash or Willie Nelson eight track into the dash. I didn't know the lyrics yet, but I would hum along with him to "Ring of Fire" and "Man in Black." I still do.

After the Jeremy Tree, Grandpa planted a maple out back for my brother Bryan and another for our cousin Jennifer. Then he apparently gave up when twelve other cousins appeared in a matter of a couple of years.

Eventually I was able to climb the Jeremy Tree when both the sapling and I were strong enough. Maybe he didn't intend it, but Grandpa was teaching us that trees had sacred properties to those open to seeing them.

I didn't plant a sapling for either of my children, but I found a way to connect them to trees of their own.

There are two specimens in a large urban park near where they both were born that we visit on occasion. One is a gnarled cherry that rises and falls quickly to the ground in a fallen rainbow shape, then back up again with a canopy that surrounds the entire footprint like a leafy room. It made for an off-the-ground catwalk of sorts when my son was a baby. I would hold his tiny hand and let him feel the thrill of being "off the ground." We call this the Z-Tree after his name, Zion. Thirty feet away is the Say-Tree named after my daughter, Sela. The Say-Tree is a three-trunk bald cypress grouping that rises high into the sky. At its top, a red-tailed hawk is often nested and hunts the park for rabbits, much to our terror (and delight). In between the three trunks is a small vacant space my daughter would crawl into and play make believe with sticks and acorns. On a random afternoon, I will say, "Let's go visit your trees," and the kids know exactly what that means. Now teenagers, I usually include a bribe for ice cream on the way there or back, and perhaps I'll bring along a Frisbee.

Unlike my grandparents, we do not own these trees, and I wouldn't be surprised if they are known by other names to other families, but they represent a place of significance to us just because we say so. The lazy cherry has lost many limbs in the almost two decades we have been visiting it, as its arched trunk sinks into the soil more with every passing year. Like my kids, these trees were once saplings as well—new and fresh and full of optimism—but like us all, someday they'll be gone. The sycamore I hid in from my dad is gone. The oak he built our treehouse next to is gone. The Jeremy Tree planted by my Grandpa Bill—both he and the tree are gone. The Bryan Tree and my younger brother Bryan are gone too. Life is agonizingly short and offensively quick. I propose we get a redo option on our fortieth birthday—just pull the lever like in *The Price Is Right* and the year it lands on is what you get to go back to.

Sometimes I'm frustrated by the reality of aging, but then I look at the maple leaf fading from green to orange to brown in autumn, and it doesn't seem concerned. Its time is short too, and we know another leaf will replace it in the spring, bringing new color to the world.

Around the same time I was hiding in the sycamore from my dad, a teenage Anthony Ambrose was growing up in Chico, California, where he climbed the tallest valley oaks in his neighborhood. He recalled being awestruck by them as a child, then eventually falling in love with the redwoods after being introduced to them by a young college professor of his. This love story led to his master's in forestry at UC Berkeley. His thesis started with investigating landscape-scale forest dynamics, but when he took a forest canopy biology class his first semester, his eyes were opened to the dynamics of science in the canopy. Anthony quickly shifted his focus into investigating the water-holding capacity and microclimate effects of canopy soil and fern mats in a single coast redwood tree named Atlas at Prairie Creek Redwoods State Park in Northern California.

My wife, Aimee, and I met him when we signed up to volunteer and learn about one of the world's largest living organisms, the great sequoias.

Like any true field botanist, Anthony's hair was disheveled and his beard a few weeks behind a trim. He had striking blue eyes and calloused hands, and he moved in fast, precise motions like a nuthatch building a nest. When Aimee and I arrived at Calaveras Big Trees State Park, southeast of Sacramento in the Sierra Nevada, he was systematically packing kits for this season's studies of the giant sequoias. Ropes, tree-climbing gear, and various instruments I didn't recognize were spread out like a swap meet on the ground.

I motioned to the piles. "How can I help?"

Anthony's verbal response cadence was much like his movement—the fragmented chittering of a songbird. "Who, me? Uh, yeah, no, no, I don't need help. Ask Wendy. She'll give you something to do. She's over there. Glad you're here." He scratched his head and motioned toward a slender, dark-haired woman at a picnic table organizing paperwork and study equipment with other volunteers and diligently taking notes in a yellow notebook.

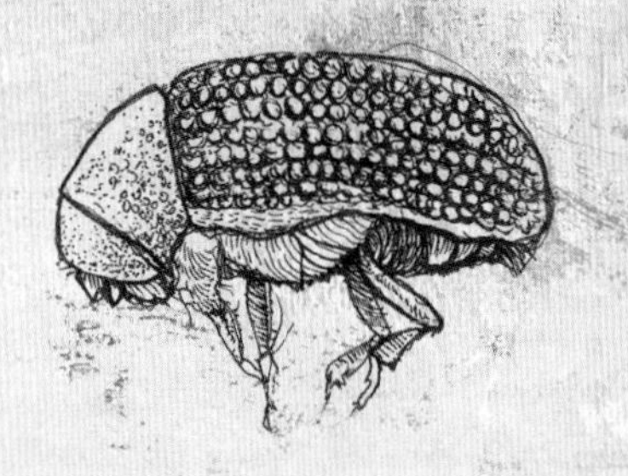

Wendy Baxter had been trying to save these great beings for over fourteen years after also graduating from UC Berkley. It was during Anthony's postdoctoral work in the redwoods that they partnered up in life and in a shared mission to learn everything possible about the biggest trees on earth and pursue saving them. They co-founded the research-based nonprofit called the Ancient Forest Society with the goal of protecting old trees and ancient forests by studying how they function and their biological diversity—especially important now as our environments are so rapidly changing.

Right out of school and despite minimal experience, Wendy landed a job with the well-established Save the Redwoods League. She credited luck and timing but also her background as a rock climber with helping her secure the early dream job. "I grew up trad climbing in the Gunks of New York. It wasn't trees, but I was comfortable being exposed and way off the ground. I don't know if that helped me get the job, but it sure didn't hurt."

This was the beginning of a long season for Wendy and Anthony's giant sequoia research. These mountains are the only place the trees grow naturally, and roughly half of the giant groves are located farther south in Sequoia and Kings Canyon National Parks. In 2020 and 2021, the giant sequoias fended off flames, smoke, and damage from massive wildfires caused by historic drought and a continually warming climate. But rampant wildfires were not their only threat. In recent years, scientists had revealed a slowly increasing impact of statewide bark beetle infestations weakening and killing the giants in true David and Goliath fashion. More often than not, this occurred as a one-two punch: the drought and severe fire damage weakened the trees, and the beetles finished the job. In the course of only a few years, the population of mature giants in California dropped from 75,000 to 60,000. If the giant sequoia were an animal, it would be at the top of the endangered species list.

PROTOCOL 2:
BARK BEETLE SURVEY
1 BRANCH PER ZONE,
THREE BRANCHES
TOTAL
PROTOCOL 3:
TERPENE. ONE BRANCH
PER ZONE, 2 REPLICATE
PROTOCOL 1:
BRANCH COUNT - ALL
BRANCHES ALONG
CLIMBING ROUTE

Wendy tasked me with preparing nine-by-four-inch strips of aluminum foil, which then got folded into small envelopes for gathering samples from the giants' canopies 250 feet above us. Samples from the foil envelopes would then be transferred into liquid nitrogen vials to take back to their lab to study. As the other volunteers tackled various preparation tasks, we got to know one another. There was JT from Humboldt, who came with his friend. He was already wearing his harness and helmet, ready for whatever came next. In fact, he never took his gear off the entire time we were there. (I wondered if he slept with his gear on.) He was in all purple—purple glasses, purple pants, and neatly tied purple shoelaces threaded into his weathered brown work boots. I noticed and asked him about it. "I just like purple."

Scott Baker, a white-haired, accomplished arborist from Seattle, held court on all things sequoia, sharing both his vast knowledge of their ecology and the myriad options for ascending them.

In fact, everyone geeked out on the tools of the trade, talking about personal experiences with various rope styles and ascending gear and techniques, as well as offering theoretical advice on different uses. They sounded like rock climbers but even nerdier. Throughout our work in the park, Scott went on to freely share his knowledge with numerous visiting school groups, his love of the forest evident in a life dedicated to knowing all there was to know about them.

A twenty-year-old volunteer with a furry mustache and quiet nature approached me. "My name is Christian, but you can call me Dibs." "Why Dibs?" I asked. "I used to call dibs on everything when I was a kid, so my dad just started calling me that. It stuck. Now everyone calls me that."

Dibs had been learning to work with redwoods for the last month and was hoping to land a forest ecology job in Oregon. We were paired up by Anthony to work together. I would be ascending free-hanging ropes with a box cutter and kit to retrieve needle samples from the highest canopies of the trees. Dibs would stay on the ground to support and send up something if I forgot it.

After Wendy and Anthony gave us step-by-step instructions for each of our assignments, the packs went on and we were off to visit our various giants and gather intel. We were sampling pine needles for moisture content and bark samples for beetle activity, but both were done up in the canopy. A crossbow loaded with a blunt-tipped arrow was shot over the highest limb visible while dragging a fishing line. The fishing line in turn pulled up a three-millimeter line before transitioning to a static ten-millimeter rope. On the ground this rope was anchored to a nearby tree, then tested for bodyweight. This process was often repeated as the first limb hooked was the lowest, at about 80 percent of the tree's overall height, and we needed to get up higher in the canopy.

On the ground, beauty was sprouting amid disaster. Massive exploded tree carcasses lay all around, covered in char and broken limbs. Splintered trunks like tombstones were scattered about the forest floor, hollowed out by time. Among the shrapnel, though, were oak and sequoia saplings poking their way through the decay, emerging from burnt remains. Joining them were groves of shy mushrooms and elusive scarlet snow plants, while various flowers peeked out from mounds of pine needles. With black-soot fingerprints on his cheeks and forehead, Anthony reminded me that fire was how these trees had survived for over three thousand years. "The giant's survival is the result of dozens, maybe hundreds, of small burns. The history of this area is based on fire suppression. It was saved because it needed saving, and it wasn't even until the 1960s that scientists figured out fire was actually a benefit."

DARK-EYED JUNCO

The bases of all the giants in the park were ash black and powdery to the touch. It was impossible to stay clean as I prepared my ascending gear to climb. Many of the trees had hollowed-out sections at their bases called fire caves that offered a peek inside. Some of these hollow cavities are as much as 50 percent of the tree's girth.

Anthony left Dibs and me to our work, and I began heading upward on a rope that disappeared into the sky. Free hanging twenty feet from the base, I quickly felt the exposure. It's not uncommon in rock climbing to be exposed, but here I hadn't even seen exactly what I was anchored to above. Without too much trouble, I reached the first limb, already one hundred feet off the ground. I radioed down to Dibs, "What's the protocol for passing a limb this big?"

He radioed back, "I have no idea. My first tree climb was seven days ago on a tree half this tall."

Although we were volunteering, I realized it was me and Dibs who were receiving the benefit. We were learning, supporting a cause we believed in, and doing something we otherwise couldn't: it's illegal to climb old-growth trees in a California state or national park unless approved for scientific research. The experience was a gift.

As I navigated through the gauntlet of limbs two hundred feet above the ground, I was able to occasionally do some moves of actual limb climbing upward while staying attached to my rope. This came naturally to me, and I giggled at the joy of it. I admired small, juvenile limb sprouts poking out of the trunk like gummy worms with tiny cones dangling from the ends. Each cone would eventually have hundreds of seeds inside it, and the only way these seeds would get released as hopeful offspring was through heat. The occasional fires in the undergrowth were crucial to the trees' survival.

At the top of my rope, I found it ran through a suspended pulley attached to wide slings to avoid any friction on the bark. I radioed down to Dibs, "I forgot my knife." Dibs tied it to the rope, and I hauled it up. It was only when I started taking samples that I recognized the absurdity of using a sharp knife next to a weighted rope. One wrong move and . . .

I slowed down and remembered my instructions: gather a sample smaller than a pencil's thickness by making a cut smaller than a quarter-inch at the fork from the main branch so that it could fit into the receiving end of a device called a pressure chamber.

Looking across the valley, I could see Wendy at my same height—a small red dot in a neighboring giant. I raised a fist and yelled, "hell yeah!" She responded with coyote calls, which I returned.

These were clearly my people—fascinated and in love with everything the natural world has to offer. It felt otherworldly and special to me. We were here just as volunteers, but Wendy and Anthony had dedicated their lives to this research. I rappelled to the ground, and we hustled the samples to an awaiting cooler to maintain integrity.

My alarm woke me at three o'clock in the morning. With perfect springtime conditions, I was sleeping in the open under the pines with a clear sky at a brisk 40 degrees Fahrenheit. I rolled out and grabbed my pack. As Anthony brewed coffee nearby under the glow of his headlamp, Wendy reminded us why we were up so early. "We are trying to measure how stressed our study trees are by collecting foliage samples," she said. "We collect them when they are the least stressed, right before the sun comes up, after they've had the evening to recover, and then again at midday when the sun is at its highest. We compare these timepoints to assess how stressed a tree is. The foliage is under tension, and when we clip our samples, the water column retracts back into the stem. We use an instrument called a pressure chamber to force the water back out of the stem, and the amount of pressure this requires is equal to the tension the foliage was under."

I nodded quietly, trying to fully understand the process, and did some light stretching before hiking back to tree #104. We passed through the silent grove, and I found the rope (fortunately) still dangling in the dark. (We had anchored the ropes out of sight of park guests to save time on re-rigging, which can take half a day.) A maze of branches were silhouetted against a dark indigo sky splattered with stars. I attached my ascenders and rose into the darkness. At the top, I again took samples and placed them in a bag attached to my harness. I clipped into my descending device and checked my system for safety. But then I stopped. Why was I in a hurry to get down? Instead, I sat on a limb and breathed quietly as sunrise crept over the northern Sierras.

I was that floppy-haired Kid in the sycamore again.

I wasn't afraid anymore, but I still felt a need for moments of hiding from the chaos of the world. Up there was a place to leave it all behind—much like what rock climbing offered me. Being off the ground introduced a radical departure from the y-axis and had felt as normal as walking to me for as long as I could remember. It was not just the trees that offered a place of safety; it was also the rhythm of leaving the earth for a moment. Birds started to chirp, and the sun rose as beautifully as a Philip Glass song—sunlight hitting every little thing like strokes on piano keys all around me. I was just another limb of the forest.

RATE
VALVE
PMS Instrument Company
Pressure Chamber Instrument
www.pmsinstrument.com

Back at camp, we each excitedly reported our predawn experiences and produced our samples for study. Without hesitation, Anthony began taking notes and documenting our deliveries. He played heavy-metal music on his stereo and disappeared into his work.

I asked Wendy about memorable interactions with wildlife up high in the giants. "Oh, mostly songbirds—nuthatches, chickadees, and wrens. Occasionally a Cali spotted owl, which is special. Then there was the time a flying squirrel leapt out over my head and floated to another nearby tree."

I chatted with fellow volunteers Rikke Næsborg and Cam Williams. They had been doing sequoia research of their own since 2005 and worked full-time at the Santa Barbara Botanic Garden with PhDs in botany. Rikke, a tall Danish woman with flowing white hair, reminded me of a youthful Jane Goodall and is one of only two lichenologists in the state. She inhaled, scanned the canopy above us, and said with a smile, "Being out here is being at home."

In the afternoon, I made time to draw, and after a while, Anthony took a break from data entry for a look over my shoulder. "Great drawing, but the needles don't work like that."

Embarrassingly, I had drawn the most generic version of pine needles, so, of course, the expert immediately saw that I was faking it. I was not observing well. I was not seeing what was right in front of me. He scurried to his mobile lab and returned with a sprig of needles. "Look, this is one that you brought down. It sprouts outwards like this and doesn't dangle. It's entirely different than any other needle." I immediately pulled out white paint and covered my mistake so I could redraw it correctly.

300
FEET ABOVE

Our team of volunteers repeated the pre-sunrise and mid-afternoon gatherings for two more days, and on our final afternoon, I saw Anthony and Wendy talking excitedly and quickly packing up their lab. They had just received confirmation of a dream specimen study, getting approval to climb and retrieve data from the largest tree in the world—the 2,200-year-old giant sequoia known as General Sherman in Sequoia National Park. This would be the first recorded effort of it being climbed, and they had to be ready to go in twenty-four hours. I asked Anthony what he expected to find on Sherman. His response was poetic and focused: "There are multiple lifetimes of questions to ask about these trees. So I continue coming out and asking."

"One of the more flexible words in the English language, ART is defined as a human creative skill, doing something as a result of knowledge and practice. It is the mind's eye rendered in visual form."

CRAIG CHILDS

TRACING TIME: SEASONS of ROCK ART on the COLORADO PLATEAU

I have stacks of sketchbooks hiding in my attic that I'll never show anyone. They are full of "mistakes," like drawing sequoia needles incorrectly. They are an archive of a young person trying to find their way through ink and graphite on paper, and I've always felt they just radiated mediocrity. But I had to get through it. All of us who pursue our voice do. We have to have compassion for those younger versions of ourselves, so intent on finding our purpose and making our marks. The pages are up there in the attic, quietly reminding me how far I've come and how far I still hope to go. To me, those sketchbooks are not full of drawings but rather the crude impact marks of a pickax digging away at the crust of who I wanted to be.

They would make for great kindling. And, I guess, in a way, they did. They were the beginnings of an inner fire of a lifelong pursuit. Decades later, the drawings in these pages were done primarily in my lap while traveling. I am neither proud nor embarrassed by them. I am only a conduit they moved through.

Ten years after Céline Cousteau invited me to join her in the Amazon, I went to the sequoias to again do what she had prompted me to do: observe, find a connection, and draw. If I had a business card, I think that's all I'd put on it, just to be cryptic. It's really all I want to do. Caring about something in need as a creative is an intoxicating concoction, but figuring out the HOW is rarely easy.

Some are fortunate to find a life path when they are young. In my case, I knew from an early age I wanted to be an artist, whether or not I knew entirely what that meant. Like any ten-year-old, I would replicate cartoons crudely on blue-lined notebooks, then fell in love with drawing horses and, of course, trees. I was never the best in my classes technically, but I was and still am entirely driven to make marks with every fiber of my being. As long as I draw breath, I will draw. For better or worse, I walked through the door of the first university that offered me a scholarship, but in the end, it did nothing but slow me down and steal my money.

I tried everything—digital art, collage, painting, sculpture, and so on—but my true love has always been the simple act of drawing. As I contemplated writing another book, I looked back through my travel journals, and the most consistent totemic symbols were trees. I had to know what was behind that.

Is there anything more natural than drawing a tree?

Hand anyone a crayon, from toddler to grandparent, and they can replicate what they know as a tree in passing fashion. Depicting a pine is as easy as a triangle with a stick. However, when looking at humans' earliest markings, plants were not a common icon. Our ancestor artists seemed focused on animals, people, and patterns. Animals brought reverence and sustenance. Patterns indicated possibly a tribal pride, or, who knows, maybe they just enjoyed making marks. Plants seemed a low priority. One of the earliest-known visual expressions of a tree, however, is remarkably presented as a deity.

Over twenty-five thousand years ago in Brazil's Serra da Capivara National Park, artists left hundreds of cave drawings for us to contemplate. One of the more fascinating images is that of a tall tree being worshipped by a group of humans. Found in 1970, it is considered to be the only example of rock art over ten thousand years old depicting a plant. The faithful raise their hands in worship in a radius around the trunk. It's safe to say the congregants were a group of men (all are depicted with a protruding "third leg"). Perhaps this is when the great Brazilian samaúma tree first became sacred, but to draw anything is an opportunity to make it sacred.

Like the earliest people, we can draw to understand, to celebrate, or simply to enjoy the act. There is a magic to making art, but like any street magician, tricks can be taught. Drawing or any creative act should not be relegated to those who are seeking to make a living from it. Creating is for anyone. Like singing in the shower or beating a bongo drum, drawing for the sheer joy of it is something far too many abandon in elementary school.

We tend to turn masters into deities and then they seem unreachable to us on their pedestals. When people say, "I can't even draw a stick figure," it feels like nails on a chalkboard to me. This mindset comes from the grade-school position that dictates who is "good" at a thing and who isn't. The joy of making "just because" is for any of us. Being an artist isn't strictly about producing something of quality; it can also be about experiencing the quality of something.

In retirement from engineering, my dad has taken on puzzles, gardening, and elaborate meat smoking. There is no quarterly report card waiting for him. He's simply seeking to find joy in the process of learning and making, not for anyone's approval or comparison, unless, of course there's a local chili cook-off—then all bets are off.

EPILOGUE

May the roots of
suffering diminish.
May warfare, violence,
neglect, indifference, and
addictions also decrease....
May we continue to open
our hearts and minds,
in order to work
ceaselessly for the
benefit of all beings.
May we go to the places
that scare us.

Pema Chödrön, The Places That Scare You

I draw millions of trees a year but not as individual arboreal portraits; rather, most often they are small, repetitive blobs from above on aerial maps indicating a forest. There's a rhythmic meditation to it. I know intimately the erratic, loose-wrist motion required for deciduous species and the appropriate pressure to apply to indicate a massive evergreen grove. For small shrubs, a minor scribble works with a hint of a suggested shadow. Imagine signing your signature but staying in one place on the paper. The motion for indicating tufts of desert grass is like throwing a tiny Frisbee without moving your arm.

Despite developing a limited understanding via repetition, I've never enjoyed teaching technique. I prefer pushing thought processes. In a short three-semester stint as an adjunct professor at an art institute, the name of my class was "Conceptual Problem Solving." I drove students crazy with assignments that expected them to come up with strong concepts, but did not require them to complete the work. "This would be great for my portfolio if I finished it," they'd complain. "Great," I told them. "Finish it, but not for a grade. I'm interested in improving how you think, not how you draw." That was not to propose that a visual skill set is without merit. One needs at least some skill to manifest what they are trying to present to the world. If your visual language is so far-fetched no one can comprehend the message, then we are just white noise. Concept will always be king.

Robert Henri, the great painter and teacher from the early twentieth century, told us, "I do not want to see how skillful you are—I am not interested in your skill. I want to know—what do you get out of nature? Why do you paint this subject? What is life to you? What reasons and great places have you found? What excitement, what pleasure do you get out of it? Your skill is the thing of least interest to me." Henri's book, The Art Spirit, from 1923, is one of my bibles. I thumb through it on occasion, looking for spiritual enlightenment.

He went on to say, "The technique learned without a purpose is a formula which when used, knocks the life out of any ideas to which it is applied." There's a balance, then, of NOT caring about the quality of the image and being equally cognizant of its necessary means of communicating.

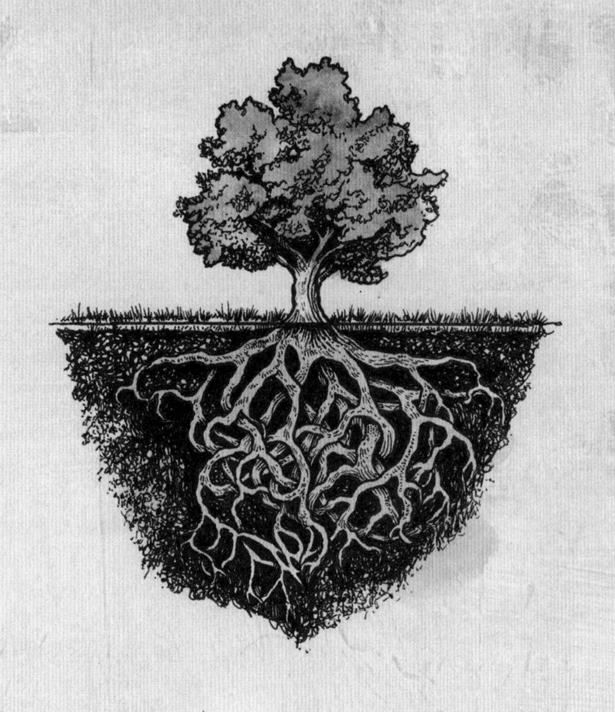

For a few years, I joined a team of instructing artists for Legendeer, an adult summer camp for artists. We visited Yosemite, Zion, Banff, Jasper, and New River Gorge National Parks in groups of fifty. Each group hosted nightly instructors and daily activities like climbing, rafting, and hiking, all with our sketchbooks and cameras. I joined one of my art heroes, Sterling Hundley, on his vision for these gatherings and found a real purpose in guiding aspiring artists on another step in their journey toward knowing themselves. In turn, this helped us instructors do the same as well.

On one of these occasions, we were fortunate to have the great illustrator Barron Storey join us. His journals tumbled out of a rolling suitcase like gold bricks onto a table. There were hundreds of them—chaotic and grotesque and divine all at the same time. Barron breathes drawing as naturally as air. His work is music caught on paper. I was in utter awe of his draftsmanship and vulnerability, but more than anything, I admired his productivity. Barron became known in the 1970s for book covers like Lord of the Flies, Fahrenheit 451, and eventually the Sandman comic book series. He has become a hugely influential instructor for generations of artists seeking to tap into purpose.

I don't use this word lightly, but Barron Storey is a legend.

At Legendeer, Barron explained that our sketchbooks and journals were far more than places to practice; they had the potential to be our most honest and profound work whether writing, drawing, or experimenting. The idea of SEEING as a practice began to integrate into my understanding of drawing on location.

Barron echoed many of Robert Henri's thoughts, but in more of a signature Barron punk-rock style:

"Nobody really gives a damn whether you draw well. Maybe your friends will compliment you. Maybe you will get an award. So what? People are living lives that have questions. If your work has answers, then that is what they need. Is being rich and famous important? No. What is important is to serve."

If you are someone who aspires to draw well, read that quote again. If not, replace the word draw above with literally anything. Garden. Climb. Write. Teach. Ride a bike. This approach has changed me—seeing that the little drawings I make somehow serve a purpose. How preposterous is that? I realize this also potentially contradicts the encouragement to draw just for the enjoyment of it, but it all comes back to your intent.

What do you care about? What are you concerned about? Pick up your pencil and start there. Every good idea starts with a pencil. A tree is just a tree until we connect with it in a personal way, and for me, that means to go experience the tree in person. To feel its bark, to sit in its shade or branches, to tap the prickly thorns of a locust and know how sharp they really are. How tall is it? Does it have its seasonal leaves or not? Do the needles dangle straight down or extend in all directions? A Joshua tree isn't a saguaro isn't a sequoia. They are all different, and without visiting them, how can we truly know? If I described a Joshua tree to you without a visual example, you'd think I was crazy. What a ridiculous tree.

To understand the Matis and Marubo and their needs, Céline went directly to them. To respond to the earthquake, the Messengers went where the need was, rather than leave and launch a fundraiser. The Ancient Forest Society would have vastly less impact if they weren't willing to go to the trees they intended to protect. It is the responsibility of a creator to KNOW about the subject they are creating. Creating on location is for anyone to enjoy, but you have to GO THERE to do it. So, go.

Pairing process (creating) with love (caring) is a powerful way to draw.

If we love a vulnerable thing, or place, or creature we are more likely to go to great lengths to protect and connect with it.

I'll ask again: What are you concerned about?

If you are eager to create (or support) work with impact, start there. I recognize and am grateful I have been fortunate to travel, but impact via creativity can be made anywhere. Our desks. Our communities. Even in the dark forest of our own minds.

In his book Knowing the Trees, environmental educator Ken Keffer writes, "After a fire, a landscape can look bleak. An unfathomable amount of forest biomass has gone up in smoke, with only a few charred timbers standing tall, like gnarled, blackened sentinels looking out over the disturbance. Yet, for many trees, this natural process is key to releasing the seeds that will become the next generation. Species such as lodgepole pine, Jack pine, and giant sequoia have specialized serotinous cones that require the heat of a fire to release their seeds."

The fact that giants grow at all is nothing short of a miracle. The chicken-egg-sized cones must first fall to the earth, then survive either fire or animals chewing through their robust shells. THEN one of hundreds of fingernail-sized seeds MIGHT survive long enough to start their journey upward as their roots twist into the damp earth. Every mature giant sequoia is a magic trick by nature. When I think back to the lessons Wendy and Anthony taught us beneath the giants, perhaps the most striking is the astonishing fact that not only do they survive the occasional fire, but they also THRIVE because of it. Giant sequoias reach full height when they are between 500 and 750 years old, and surviving difficulties is paramount to their success; in fact, without the trials, they would never grow to their full potential nor have enough exposed soil to produce future offspring saplings.

I'm sitting here shadowed by six 80-foot-tall giant sequoias. They are barely toddlers in the lifespan of their species. Crowned with dark green foliage and a soft cinnamon-colored bark, they are all close enough that their branches intermingle like ancient folded fingers. As symbols of resilience, they have withstood drought, blizzards, and earthquakes.

Barely a sliver of visible daylight lands on the dirt down below them, and the pine aroma is enchanting. They are the real thing—what commercial scented candles try to replicate but never can. Sitting in their shade is a lovely way to end the day, but I'm not with scientists or activists. I'm not on a journey sleeping in a boat on the Amazon or a truck bed in the Annapurna range. I'm not even in a national park. I'm just in our yard with Aimee and Rooster the dog in the San Bernardino Mountains of California.

We are fortunate to live among giants.

When you look at an exposed stump or slice of a cut tree, the concentric rings give insight to the trials and struggles any given specimen has experienced throughout a year or its lifetime. The rings, of course, show time and age, but it is the scars and irregularities that reveal what a tree has endured. Based on the color and width of the rings, the trained eye can decipher seasons of drought, the occasion of fire, or changes in climate and disease. And we can see where a tree has learned and changed course. It can be more conservative with its water usage after a drought. It can thicken its bark to survive infestation. A tree learns.

We can't see our own rings, but what we have been through in life manifests in other ways. A carpenter's hands are thick because of labor. A health scare or accident may take away appendages, but a broken spirit can whittle us away to a stump. Somewhere deep within the heartwood of us all, you can see it—the scars, mistakes, and battles we've weathered. The nicks and cuts in our rings expose what we have learned along the way.

Shortly after learning of the benefits of fire, scientists began performing prescribed burns to benefit old-growth forests' longevity. Low-intensity fires clear the forest floor and open the canopy to allow more light in. Native flowers return, seeds activate, and soil is revitalized. The trees are given another chance to thrive. That's what we are doing when we actively say yes to following new understandings of ourselves and the world we live in—we are performing personal prescribed burns and allowing more light into our lives.

BE
WHO
YOU
WANT
TO BE.

Over the course of these journeys, I personally weathered a business lawsuit, a difficult divorce, the agony of loss, night terrors, panic attacks, challenges with parenting, immense doubt, crippling debt, and the death of a number of friends. Suffering is not a stranger to any of us.

All of us find ourselves under its shadow at some point.

My bark is thicker because of it, however, and I've found, much like the giant sequoias, there is life on the other side of the fires we endure, and an opportunity to burn away anything hindering our growth to make room for what comes next.

the end

ACKNOWLEDGMENTS

A tree doesn't grow without sunshine and soil, and this book couldn't have come to BE without the support and input from some wonderful people. Thank you all so much.

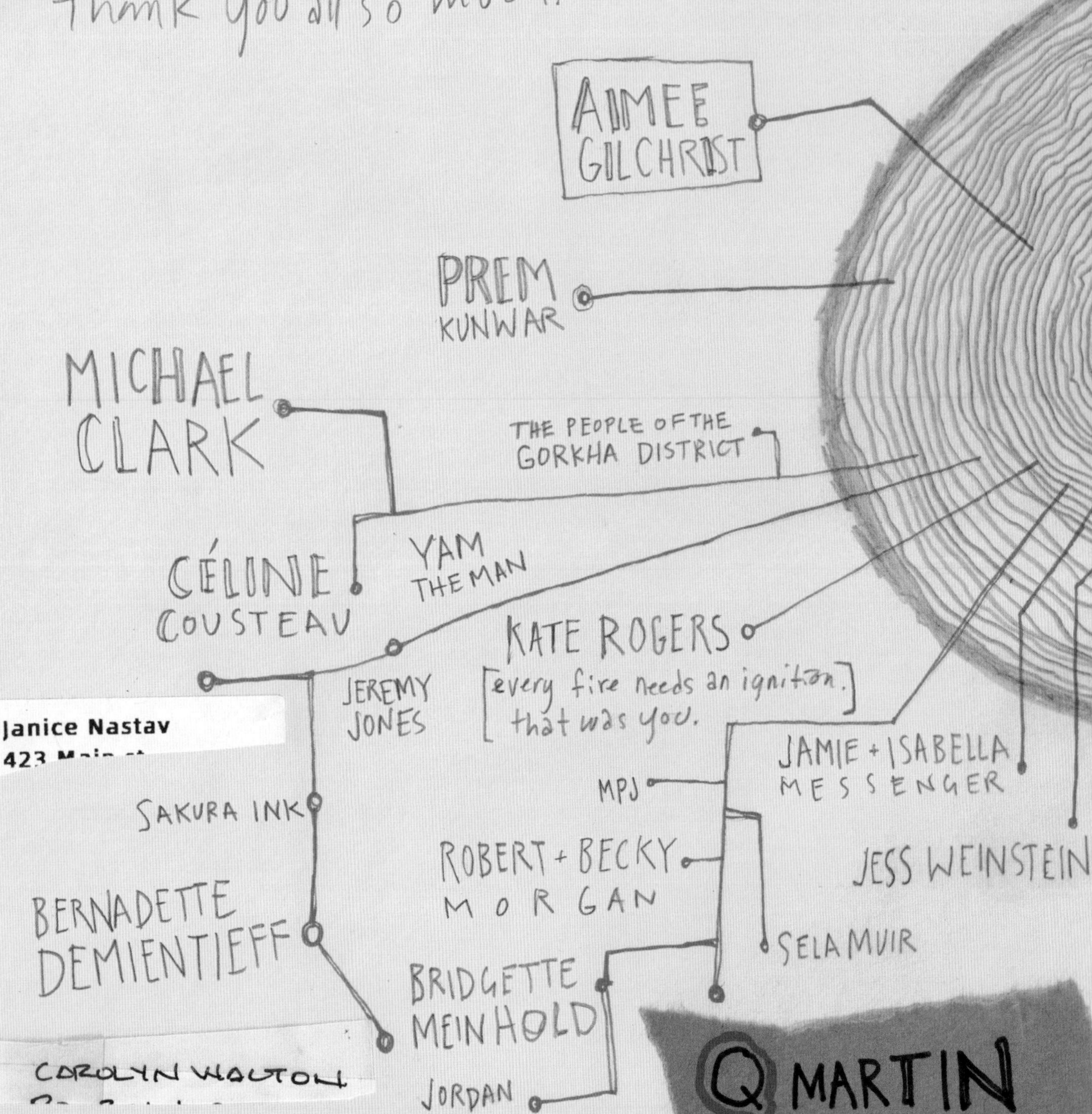

DAN RITZMAN

ANDREW JOHNSON

STACY BARE

PORCH-TIME BOYS

MATT + KRISTEN REDD

STERLING HUNDLEY

BARBARA ARISI

ROB BONDURANT

STEVEN BAUMGARDNER

DIBS

JOHN LONG

MAURY BIRDWELL

PETER ELSTNER

JEN GRABLE

BETH JUSINO

CRAIG CHILDS

NICK GREECE

JDC

RENAN OZTURK

MY BROTHER SCANNER

WENDY BAXTER

ANTHONY AMBROSE

Borja Antolín Tomás

THE PEOPLE OF THE JAVARI

FRANKLIN FRANZEN

PETER Goffstein

ZION RAY

RYAN RIMMER

GEORGE SAMPLE

THE GWICH'IN COMMUNITY

Photo by Chase Castor

ABOUT THE AUTHOR

Jeremy Collins is a multifaceted artist, climber, and adventurer whose work intertwines the grit of the wild with a mastery of visual storytelling. Known for his breathtaking illustrations and striking narratives, Collins captures the raw beauty and intensity of the natural world through his art and prose. As a climber, his daring ascents in some of the world's most remote and challenging locations have shaped both his work and worldview, fostering a deep connection between him and the rugged landscapes he immerses into. Collins' work invites us to understand the challenges and struggle of wilderness and ourselves, sparking a deeper appreciation for both the environment and the human spirit's capacity to endure.

His complex, cerebral, and whimsical drawings and maps have been featured in books, films, and commercial work. From the cover of National Geographic to his award-winning book Drawn: The Art of Ascent and his illustrations in Earth Almanac, Jeremy is a bottomless well of inspired expression.

recreation · lifestyle · conservation

MOUNTAINEERS BOOKS, including its two imprints, Skipstone and Braided River, is a leading publisher of quality outdoor recreation, sustainability, and conservation titles. As a 501(c)(3) nonprofit, we are committed to supporting the environmental and educational goals of our organization by providing expert information on human-powered adventure, sustainable practices at home and on the trail, and preservation of wilderness.

Our publications are made possible through the generosity of donors, and through sales of more than 700 titles on outdoor recreation, sustainable lifestyle, and conservation. To donate, purchase books, or learn more, visit us online:

MOUNTAINEERS BOOKS
1001 SW Klickitat Way, Suite 201
Seattle, WA 98134
800-553-4453
mbooks@mountaineersbooks.org
www.mountaineersbooks.org

An independent nonprofit publisher since 1960

YOU MAY ALSO LIKE:

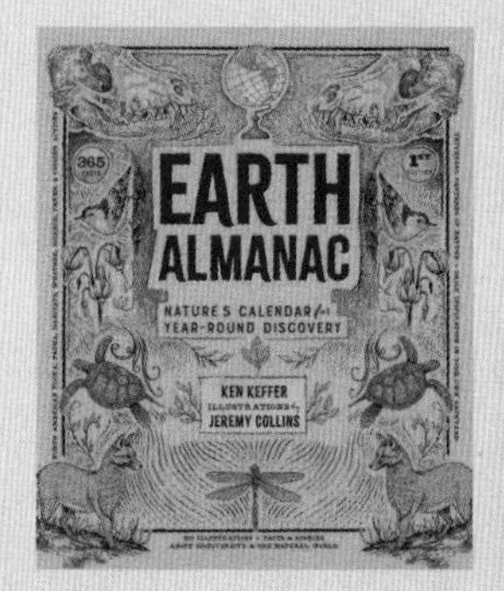

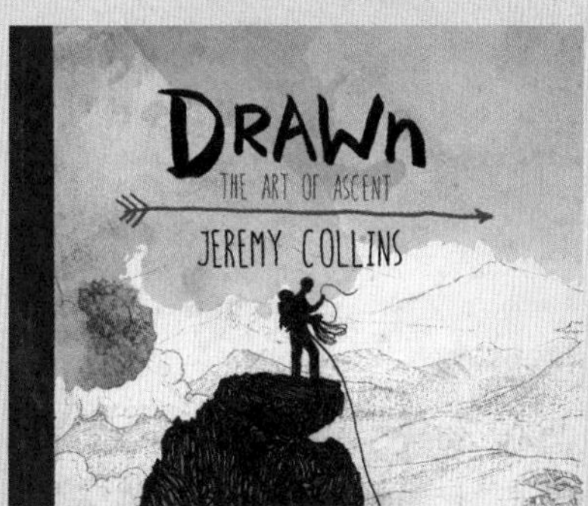

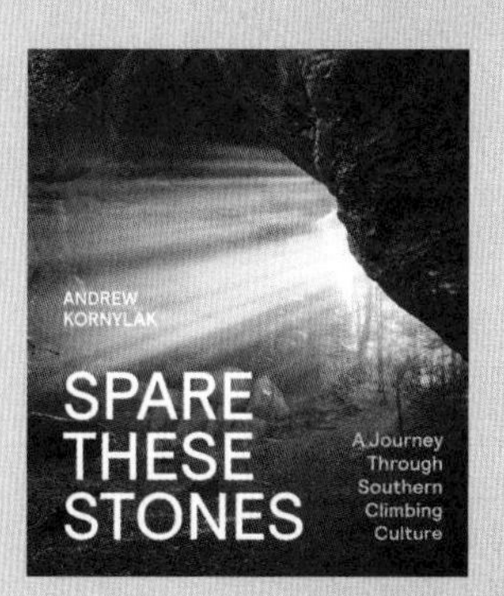

photo: Michael Paul Jones

Find me in the forest
Find me in the hollow
Find me in the far away
Where others dare not follow

•

Find the parts I've lost;
Gather up my branches.
Gather up my sticks and twigs, and
Gather second chances

•

Gather up your circle;
Your kin, both blood and chosen
Gather up those left out
Gather up the frozen

•

Gather together, 'round the flame
In the dirt far down below
For part of me I thought was dead
When gathered, still I grow.